DAUNTLESS

Leadership lessons from the frontline

RACH RANTON

Published by Change Empire Books

www.changeempire.com

Printed on demand in Australia, United States
and United Kingdom

Edited & designed by Change Empire Books

EBOOK ISBN: 978-0-6487453-4-1

PRINT ISBN: 978-0-6487453-5-8

For Damo,

Who always supports me, no matter how wild the idea or how big the plan.

"I think I'm gonna write a book this year."

"Sounds great—I bet it will be amazing. I'm sure you'll be in Oprah's Book Club."

Ignores both my complete lack of writing experience and lack of connections to famous and influential people.

I love you and thank you for always being my biggest backer.

and for Will,

There is nothing quite like a teenager for keeping your ego in check.

"Wanna come to meet the Prime Minister with me, Will?"

"No thanks, Mum. Sounds boring, take one of your friends."

Thanks for keeping me grounded and reminding me what is possible if you put in the hard work. We love you for your kind heart, giant brain and enormous work ethic.

CONTENTS

PROLOGUE

I jam the machine gun into my hip as I slog through the deep, dry sand.

Ballistic plates inserted into a green and black vest thump against my chest with each step as I make my way along the beach towards my teammates. I squint as the midday sun glares off the water and try to remember the last time I wasn't covered in sweat. The Australian Army has only been in East Timor for a few days but the steamy week spent in Darwin before we deployed bleeds the whole experience into one endlessly hot September.

The steady, rhythmic crunch of shovels in sand greets me as I get closer. If it wasn't such hard labour there would be a beauty to the monotony and assembly-line-like execution of soldiers filling sandbags.

"How many more do you think we need?" calls Jonno as I approach the group.

"Maybe sixty?" I say.

The heavy-duty bindings on the ballistic vest are pulled as tight as they can go in an attempt to make the large size gear work on my small size frame, and sharp edges scratch at the exposed skin on my neck. My salty skin makes the cuts sting.

I sweat from pores I didn't even know I had, and perspiration runs in a river down my back underneath the ill-fitting vest.

Moisture seeps into the waistband of my pants and a single drip threads its way through my bum crack and down my leg. The air tastes thick with humidity and the smell of unwashed humans.

A few relieved faces look up from their shovelling. The Squadron Sergeant Major ambles up behind me. "More like six hundred," he says. Soldiers groan.

"Six-fucking-hundred?!" says Jonno. He can always be relied upon to say what everyone is thinking.

The Sergeant Major rolls up his sleeves and takes a shovel from one of the team, gesturing for the soldier to take a break in the shade.

"That'll get us to chest height," he says. "We'll be done by Tuesday."

Our squadron returns to digging, filling, and moving sandbags. Knowing today's target—somewhere around six hundred more bags—makes it easier to push through the work.

I sling my weapon and push the gun behind my back. Linked rounds clink together as it settles to rest across my body with the butt at my shoulder blade and the barrel tapping behind my knee.

Jonno's camo shirt is soaked to his skin as he passes me a full sandbag. He stops digging, resting his shovel against his legs. He wipes the back of his hand across his forehead and flicks a sheet of sweat into the sand.

"How hot is it, Sir?" Jonno says with a grin, needling for banter.

"It's so hot, I saw two trees fighting over a dog," says the Sergeant Major.

"It's so hot, I saw a dog chasing a cat and they were both walking," shoots back Jonno.

"It's so hot the bread at the shops is already toast."

The other soldiers raise their voices in a chorus to complain about the dad jokes.

Jonno laughs and holds open a hessian bag for the Sergeant Major to fill while I slot into a human chain and shift a sandbag to the next person in the long line leading to our truck.

Teammates stand guard. The rest of us fill sandbags.

Despite the irreverent chat, there is tension. Elements of the Indonesian Army remain within the city, and we are unsure how they will react to our prolonged stay and intention to deploy across the country in support of East Timorese independence. Those standing guard watch the Indonesian soldiers, who in turn watch us.

We rotate jobs with unspoken gestures and as the machine gun bangs around on my back I dig deeper into wet, heavy sand and top off a sandbag. As my teammate ties the hessian into a neat knot, I hear a murmur run through the team.

"That's six hundred," shouts Jonno.

When the final sandbag is stacked on the truck, a driver swings down the sides of the tray. Twenty of us clamber in, some stepping onto the oversized 4WD tyres to scramble up and perch on the pile. I attempt to climb onto the tyre but my short legs won't let me reach. I climb a ladder at the back instead.

We travel back to the city with the canvas rolled up sitting high on hundreds of sandbags. The hessian scratches at my legs through my cam pants and threads of the coarse fabric waft in the air. Jonno sneezes next to me as we bump over potholes and I watch the city flick by over my gunsight. The oppressive Dili air is heavy with the smell of diesel, burning rubber, and rot.

We are fortifying a compound full of bare concrete and roofless buildings. Every structure bears the scars of fire, violence, and destruction. The sad eyes of the locals hint at the scars they bear too.

"I've got a job for you," says my boss as he passes me once we arrive in the compound.

I am suspicious but I say nothing. Senior soldiers and officers distribute regular work with a simple statement, but crap jobs tend to come with a preamble like this. You can't let the seniors know you're onto their suspicious-crap-task distribution methods and, as a lowly signaller, it is redundant anyway. My rank does not afford me the opportunity to say no.

"Yes, sir," I say and follow him to the Head Quarters tent.

At the camouflage netting I dip and slip like a boxer avoiding a punch combination to get through the tight opening. I have to stop and untangle my weapon from a snag before I get to him. He gestures at a map on the wall with pins stuck into locations throughout East Timor.

"We're going in with the Kiwis to secure Suai airfield and relieve the Gurkhas and British Special Forces. We're going to send a third EW operator and they need more machine gunners."

That's what I do in the Army—I'm an Electronic Warfare (EW) operator. Our job is to intercept and analyse enemy communications and give advice on the battlefield. We find out how the enemy are communicating, listen to what they are saying, and provide analysis to commanders on the frontline.

My boss smiles as he points to a pin on the other side of the island, close to the border with West Papua.

"Tomorrow, you'll be rehearsing the Suai insertion with Victor company from the Kiwi infantry."

This doesn't sound like a crap job, it sounds like a great job. I'll be on operations, in the field, delivering EW in a real situation, not just practicing on exercise. Like most twenty-year-olds, I try to hide my emotions. This opportunity is the most exciting and terrifying thing I have ever heard and I hum with the energy of both.

The Australian Army has not been deployed on a large scale in active combat since Vietnam. When I joined the Army I had expected my own service to be fairly benign. Part of me had been disappointed but the other part had been relieved I would never have to find out if I was the coward I always suspected I was.

Now it looks like I'd be on the frontline within days.

Embedded with the New Zealand infantry.

Inserting via helicopter into a vulnerable airfield.

Carrying the gun.

All before my twenty-first birthday.

I worry I'm not good enough and tell my boss as much. He puts a hand on my shoulder.

"We could pick who we wanted; we picked you."

I keep a straight face, belying the turmoil I'm feeling beneath the surface.

"There's been a lot of militia in Suai and if things get hot when you land you'll be the first Australian Army female in combat."

"Thanks, boss," I say, swallowing.

I'm not sure I can do it, but I say yes.

After only two years in the Army, I've learned to say yes to opportunities. To say yes when others believe in me, even if I doubt myself. Already the Army has shown me many times my limits are way beyond where I think they are.

We meet the New Zealand Infantry company commander the next morning, join his team of about a hundred at lunchtime and rehearse with Black Hawk helicopter crews in the afternoon. Each team member is focused on their role and their tasks. We rehearse until we can complete every scenario and sequence without error. The tropical afternoon heat is relentless as we rehearse.

Exfil from helos.

Secure and hold the airfield.

Assault the buildings in teams.

I've never done any of it before—everything is foreign to me—I work hard to hold my own and get it right. I'm joining a team with two EW guys who are top operators. I know I'll need to be at my best to physically and mentally keep up with them.

We rehearse scenarios landing under attack, with snipers, if the helo crashes.

We rehearse completing the mission, regardless of the changes thrown at us.

We rehearse scenarios with mass casualties for us and them.

Feedback is direct and brutal. Trust is built through hard work and a shared purpose.

By dinner that same day, despite the different patterns on our uniforms we have been completely integrated and embraced by the Kiwi infantry platoon we are embedded with. We laugh, jibe, and shout profanities across the mess tent at each other.

〜〜

I'm given permission to call my parents.

"Hello, Dad?"

The line crackles and the white noise is loud in the extended pause before I hear his voice.

"Rach?"

"We're on a satellite phone Dad, there'll be a big delay, so try and speak in a block.'

Heeding my own advice, I take a breath and ready myself to say everything I need to all at once.

"I've just got a few minutes. I've been sent to East Timor. I've just arrived in Dili, and they're letting me call you because I'm about to fly out with the Kiwis—they need an extra machine gunner—the boss just told me I might end up being the first Australian woman on the frontline!" I blurt out.

The pause until I hear his voice again is even more extended this time.

"What?"

I roll my eyes.

"I'm in East Timor, Dad. They say I might be the first Australian Army female on the frontline of a combat zone," I repeat.

I wait for the sat phone lag.

I can't believe he's not more excited.

"Um, that's ... um ... good?" he says.

I grit my teeth and shake my head. I expected him to be more enthusiastic.

"Okay, well ... can you tell Mum? I've got to go."

I wait for the lag.

"WHAT?"

"I've got to go, Dad."

"OK. Don't do anything crazy. We're already impressed. Stay safe."

I roll my eyes again and hand the sat phone back to the Sergeant Major.

He looks amused.

"You might feel differently about that conversation when you have kids of your own, Rach," he says to me.

~~~

In those first few days in East Timor I see how tall the leaders around me stood. Jonno, the Sergeant Major, my boss who assigned me to an infantry company, the Kiwi company commander; leaders standing tall.

Leaders who encouraged me to stretch myself to find out what I could really achieve. Leaders who backed me to get a complex job done. Leaders who helped me quickly become one of the gang. Leaders who gave me opportunities in the years prior to that moment, so I was ready to take on any challenge.

Those people who focused on the team and spent their time and energy on building a culture of trust. Our teammates who made sure we kept our sense of humour and set clear goals. People who had the courage to do things differently and be themselves, regardless of how serious the situation.

Not all of these leaders were my direct boss or held a high rank. Some were middle managers and others were my peers who held no official title. Regardless of rank, I saw the leaders around us have a profound impact on me and our team.

I believe our leaders leave an indelible mark upon us.

I have felt the words of my leaders echoing through my life for years after they were said, their actions still inspire me decades later. The best of them lifted me up, encouraged me to be my best and made me feel like I could achieve anything.

They were not all perfect. I also remember the leaders I hated working for and the ones who made me feel small, excluded, and worthless. I've spent lots of time thinking about which ones I want to be like. And which ones I don't.

After a decade in the Australian Army, I learned to recognise a pattern of leadership behaviours I knew would bring out the best in me. The same themes, the same behaviours—pursued rigorously by the best leaders—they kept emerging. When I moved to a corporate career, I found the best leaders there did these things too.

They build great CULTURE

From nothing, from groups of strangers, with little time.

They know teams matter when it comes to the crunch, so they put all their effort into building team culture and an environment where people feel safe to speak their mind.

They tell the truth—ugly, direct, uncomfortable, or confronting.

They communicate with clarity and trust their people to do their jobs well.

They give their power away.

They focus on TEAMS

They encourage different thinking and understand inclusion is the key to harnessing diversity of experience and thought.

They help people leverage their strengths and find purpose in their work.

They encourage behaviours which align with the team's values and create opportunities for their team to put these values into practice.

They acknowledge when people do the right thing and behave in the right way.

They acknowledge when people do the wrong thing and behave in the wrong way.

They are right there in the thick of it with their team—sharing experiences, rolling up their sleeves, and providing support when things get hard.

They show COURAGE

They do what is right, not what is easy.

They embrace risk and have a growth mindset.

They back their team and they lift others up.

They are authentically themselves—the good, the bad and the ugly.

They are the leaders who take a dogged, determined, courageous approach to serve those they led. They lead with purpose and with a passion for their people. They are not fearless, but they do not shy from facing their fears. They are authentically themselves, unafraid to share their failures or weaknesses. They are bold, take risks and are confident to make decisions, guided always by their values.

They are Dauntless leaders.

Dauntless leaders succeed on the battlefield, in the boardroom, and in business.

Before you read on

While I've tried to remain true to stories as they happened, in some cases I've had to change names, combine experiences and omit some details to protect the innocent.

There is some deliberate blurring of characters as I don't want anyone to be able to claim royalties or sue for defamation depending on how I've portrayed them. If you've worked with me, you might recognise yourself in a turn of phrase or a particular scenario. If it's flattering, please feel free to claim it. If it's not, I've avoided real names so you can just pretend it's not you.

Expect to jump between different times in my life. You'll move from my final moments in the military, to deployments within my first few years, and then walk beside me on the base in a random order. This is not a memoir; it is not my story from birth to death. It is the stories of the Dauntless leaders I have worked with and the lessons I learnt from my time in the military.

There's also some details I've had to change/omit due to security requirements, so we all have to be content with dancing around the edges of some national secrets.

My family all assure me I am not funny.

I disagree wholeheartedly, but in keeping a growth-mindset about feedback, I'll take their advice and try to keep things on the straight and narrow most of the time.

Feel free to laugh at me though. While my jokes don't always hit the mark, I think you'll find some of my failures smirk-worthy.

I acknowledge not everyone's military experience is positive and many of my colleagues suffer ongoing trauma from their service.

I thank all who I served with, and who have served our nation. They say it is the people you miss when you leave the Army and this has certainly been the case for me. The larrikins, hijinks, and bonds of shared experience; I believe most veterans will remember these elements fondly.

These are my experiences and told as I remember them. I was not in any of these situations alone, and I acknowledge others may remember experiences, events, and emotions differently to how I recall them.

The stories I tell in this book touch on my time on operations in East Timor in 1999 and Afghanistan in 2006 along with a range of military experiences. Where I can I've avoided talking about specific incidences of direct conflict and instead have focused on the leadership I experienced which prepared us for those situations. If you find the content triggering or confronting, there are services I'd encourage you to engage with:

- Open Arms—Veterans and Families Counselling (1800 011 046)

- Beyond Blue (1300 224 636)

- Lifeline Australia (13 11 14)

CULTURE

Dauntless leaders focus on culture first because they know teams matter when it comes to the crunch.

It's not Taco Tuesdays and ping-pong tables that make a great company culture.

It is the people, the leadership and how everyone goes about doing their work which makes a workplace 'feel' great.

Dauntless leaders spend their time thinking about and acting to shape the culture of their teams. They know how they behave is watched by all, and they know the behaviours they endorse, accept, or tolerate all have an impact on 'how it feels to work here'.

Dauntless leaders build culture from the ground up. **Safety**, **honesty** and **trust** form the foundations of culture in Dauntless leadership.

CULTURE—before you read on ...

REFLECT

Think about the best team you have ever worked in:

- Why did you choose this team as your 'best'?

- How would you describe the culture of that team?

- Where do you think your current team is now in terms of culture?

- What do the words safety, trust and honesty mean to you?

GET CURIOUS

Ask three people in your team about culture:

- What are three words they would use to describe your current team culture?

- What three words would they use to describe the best team culture they've ever worked in?

- What do they believe your team does to build culture?

- How honest and trusting do they think people are on your team?

TAKE ACTION

Record your reflections and the responses of others.

Read the chapters on Culture—at the end you'll use this exercise to help you build a plan to shape culture.

SAFETY
[THE STORY]

Dauntless leaders build team culture by creating an environment where people feel safe to speak their mind.

In the depths of an Afghani winter, our eight armoured Bushmaster vehicles pull up in a saddle between two mountains.

We'd arrived at the end of Summer but the mirage-inducing, sand-storming, hot, dry weather of those first months is now a distant memory. By the middle of our tour it is snowing almost daily. When I'd thought about deploying to Afghanistan I didn't think about the freezing winters.

Treeless saw-toothed mountains surge above the long valley we patrol. Where there is no snow the land is all shades of brown: silty rivers and mud-walled compounds.

Vehicles halt in a staggered line spearing left and right off a track to secure the ground. Turret gunners stand tall out of cupolas, machine guns face out. The final Bushmaster swings in a big hook turn and parks to protect the road we have driven down. The Aussie-made mine-protected vehicles have already proved their worth against Improvised Explosive Devices (IEDs) in the rugged terrain of southern Afghanistan.

My gloved fingers reach for my weapon as I dismount from the vehicle.

Snow drifts down. Soft, weightless and fluffy, so unlike the wet Australian snow I am used to. A flake lands on my eyelashes and I blink it away. Another turns into an icy spot on my bare cheek.

"Okay," says the combat team commander, breath pluming in the freezing air. "Night loc—what do we think?"

Around the group, soldiers and officers offer their unfiltered feedback. Each listens with intent to the thoughts of others around the tight circle. A red-headed driver points out his preferred overwatch position, punctuating sentences with snaps and pops of his gum in minty-fresh exclamation points. A combat-engineer sergeant offers defensive options through chattering teeth and we all listen and think.

"No good for us, boss," I say.

All eyes turn to me.

"The groups which give us trouble in this area have a safehouse in that valley." I gesture beyond the tallest of the two peaks we are between. "We won't be able to hear them with a mountain in the way."

I watch the commander pull the coarse fabric of his *shemagh* up over his nose. It's harder to read his expression with only his eyes visible above the tan, beige and black of the mottled scarf.

"What options do you like?" he asks me.

"If everyone else wants to stay here, then we could turn electronic counter measures on for the night. Means the enemy can't set off any IEDs with mobile phones or radios, but it also means we can't hear anything they are saying or planning."

A few in the group exchange glances. The commander who is five ranks above me simply nods and I continue.

"We could go around to the other side of the mountain—then we'll be able to get them. We can work all night to make sure we know what they are doing."

It doesn't matter at all that I'm the most junior member of the leadership group and the only woman on the hundred-strong patrol.

We move.

〰〰

Late that evening, in the cold shadow of the tallest mountain, we sit inside our vehicle and search the radio waves for activity.

Wearing big, puffy sleeping bag pants over the top of my desert camouflage cargos, I wiggle my toes inside my boots. The v-shaped hull of the hulking Bushmaster is filled with liquid to help disburse shrapnel and absorb impact if we drive over an IED. Sitting stationary in subzero temperatures I suspect my feet are resting on a huge block of ice.

I pull the headphones tight over my ears and set off a new search sequence on our equipment. Soon we find some familiar voices planning trouble.

We listen. As most of the patrol sleeps, my teammate—our interpreter—and I speak in quiet, hurried voices. We review what we know, who we think is speaking, what the voices we intercept are describing and the code words they are using.

I take my headphones off. "We need to wake the boss."

I take my weapon with me as I creep out into the night among the sleeping soldiers. I grip the commander's boot through his sleeping bag and shake it gently. He always sleeps on the ground by the back passenger-side tyre of his vehicle.

"Got news, Sir."

He sits up alert, his green eyes focused.

"They're planning on coming up the creek line to probe our position. Said they'll come in groups of twos and threes. Bringing RPGs."

We've been targeted with RPGs before. The Bushmasters are good but not always good enough to stop a well targeted rocket-propelled grenade.

He nods. "Go tell the picquet. Let's get everyone up."

The muffled sounds of clothing rustling and short, sotto voiced conversations seem sharp and loud in comparison to the noiseless, moonlit void I had stepped into when I left the vehicle.

The patrol stands to. Every soldier faces outwards. Apart from us-we stay listening.

"They're gonna hate us if we're not right, Rach," says my offsider as the minutes tick by.

We share a grimace—our job is not always an exact science.

We have to make assumptions and draw conclusions without all the information. We take calculated guesses about what the enemy is saying, which vehicles they are counting, or which valley they're using a code name for.

We've woken the entire combat team before and been wrong. But we've also made calls to stop us driving into IEDs.

It's worth the risk. We know even in the most critical of situations we are allowed to make mistakes. We want to choose the mistakes which limit loss of life.

Each gun picquet has been briefed on how we expect the enemy to present and reports start to filter in. They can see groups coming up the creek line. Night vision goggles enhance the moonlight and allow our team to watch the shapes and bodies moving through the dark in grainy greens and greys.

"They were in a big group at the bottom. They've split up now. Coming up the creek line in twos and threes," calls in one of the gunners.

"Second group has the RPGs," responds another.

"Let's make sure they know we are ready for them," says the commander.

The shout of a challenge rings out into the cold mountain air.

SAFETY
[THE LESSON]

We were ready for the enemy that night because our combat team commander did things differently from anywhere else that I'd been in the Army.

Prior to Afghanistan, my patrol experiences were all about executing someone else's plan. High-ranking officers designed plans and issued commands. Orders came from the top—from commander to subordinates—starting with the opening sentence: "These are my orders, no questions until the end."

Commanders gave commands, subordinates followed them.

In Afghanistan, our combat team commander flipped this on its head.

We were ready that night because the people on the patrol were involved in building the plan, not just executing someone else's. On this modern battlefield the combat team commander threw away 'the way we've always done things' and instead invited the people and groups who would bring a broader, more diverse input into the conversation.

EW was not the only specialist element invited into the inner circle in Afghanistan. Combat-engineers provided new demolition and minesweeping capabilities and knowledge. IED detection, IED jamming, and IED sniffer dogs—all elements previously in the background—were now brought to the fore to play a critical role in ensuring the safety of our troops. The Commander wanted the ideas from people on the ground to shape how the plan was built. It was a courageously alternative way of leading.

> Dauntless leaders are not afraid to try new ideas and new ways of working.

We were ready that night because the combat team commander had created a culture among the team where it was safe to speak up. He'd fostered an environment where we were expected to offer our alternative views with disparate and competing priorities. We were ready because we'd been working together in this different way for months. Respectful challenge was normal for us and we'd seen the success which came when we all spoke up and shared our ideas.

An environment where we all felt safe to speak out didn't come about by accident. The commander took deliberate steps to make sure all voices were heard by setting three rules for communication in our team:

Get experts in the room, regardless of rank

In the room, in the meeting, in the briefing—whatever it was—the leader didn't want to speak to just the most senior person in a section, he wanted the experts.

He gathered the people who had the most patrol experience, who had spotted previous challenges, who had adapted their trade in the field.

Rank didn't matter, expertise did.

Speak up, challenge, and identify potential issues

The opposite of the old 'no questions' rules, the commander encouraged open debate (regardless of rank) and did so in a systematic way.

Juniors soldiers and experts would be called on to share their thoughts first—before senior leaders were encouraged to speak—so the raw, uninfluenced ideas of those on the frontline started the conversation.

Rather than seeking agreement or working towards consensus early, the entire first half of meetings and briefings were used to encourage challenge. Element experts were encouraged to challenge each other, ask uncomfortable questions about capability, and to critique post-activity reviews.

Agreement on the way forward was only on the agenda after all the potential issues were raised, discussed, and pulled apart in detail.

Accept risk and embrace errors

Admitting to mistakes, challenging ideas, and sharing weaknesses was seen as critical to the mission's success. Very different to the bravado you would expect from the military; admitting vulnerabilities was held up as a strength.

Rather than apportioning blame, mistakes were valued for their ability to provide fresh insight into enemy actions and reactions, along with tactical changes we needed to consider. Bringing a mistake to the table was valued higher than bringing a success.

If we didn't know about gaps and previous errors before building the patrol plan, they would become exposed in the worst possible circumstances. Hiding gaps could be the difference between life and death on the battlefield. No-one wanted to find out you couldn't deliver on what you promised once the bullets started flying.

In communicating this way, the importance of rank was diminished in comparison to the value someone could bring. I saw my teammates and leaders of every rank express vulnerabilities, share conflicting views, and deliver exceptional, innovative work during our time in Afghanistan under this model.

I want to say this again:

- Expertise mattered more than rank.

- People expressed vulnerabilities.

- Conflicting views were welcomed.

Therefore:

- Rank became less important.

- Innovation flourished.

Is that how you expected to hear military leadership and culture and communication described?

With these changes, the commander had created an environment of psychological safety. A space where everyone in the team felt safe to challenge ideas and offer diverse opinions.

The conflict I'd previously experienced when handling contradictory views did not exist and instead, we worked together with an accepted degree of tension which held no personal malice. He had demonstrated that he welcomed being questioned and set the expectation with all his subordinates that being questioned didn't diminish your leadership and could enhance it.

> Dauntless leaders are not afraid to be challenged.

Those who want to lead for innovation and inclusion welcome ideas even if they are in direct opposition of their own, but it takes work to keep all the players on your team comfortable with this balance between challenge and conflict.

Everyone on a team must feel safe to be themselves. They must feel safe to ask questions without fear of ridicule and to make mistakes which are not viewed as failures. Individuals need to feel worthy of their opinion being heard. It takes work to make people feel like this and to keep them feeling like this.

Dauntless leaders are vigilant about safety and spend their time working on and nurturing the psychological safety within their teams.

Turns out this type of environment doesn't just work in the valleys of Afghanistan.

Teams where members communicate equally, feel safe to speak up and openly share their mistakes are the top performers at Google too.

In 2015, Julia Rozovsky, an analyst at Google People Operations (what they call HR), led a team of researchers at Google for two years, searching for the factors which made up the top performing Google teams. They interviewed over 200 employees and looked at more than 250 attributes, expecting to pinpoint a combination of technical skills and personality needed for a Google team to succeed. They were expecting a team equation looking something like—one technical expert, three introverts and a project manager with strong critical thinking—but that's not what they found.

What the research team discovered instead was **how** a team works together matters *much more* than **who** is on a team.

The way people communicated and interacted with each other determined whether a team would be successful regardless of the make-up of the players on a team.

It didn't matter how many technical experts or high-ranking officers we had in Afghanistan, it was **how we were interacting** with each other that was making the difference. Expertise matters, but the ability to draw out that expertise is where you get true success. The combat team commander did not change **who was on our team**, but when he changed the way information was shared in our team, it changed **how our team worked together.**

> When teams feel safe to take risks, they outperform their peers.

When people feel confident no-one will embarrass or punish anyone else for admitting a mistake or sharing a divergent idea, team effectiveness improves.

It makes sense, doesn't it? When we are worried about how someone else might perceive our competence, we're less inclined to ask for clarification. When we're concerned we'll be punished for a mistake, we avoid risk and hide errors. When we don't feel safe to do those things, the ability of our team to grow and learn is limited.

When we are confident our teammates value what we have to say and don't question our competence if we ask questions, we're more able to explore curiosities, to innovate and to find creative approaches to problems. We get a buzz of satisfaction and ownership when we create a solution, and when we work in teams where we feel our ideas are valued, we're more likely to offer them up.

That's what was happening in the combat team. Our ideas were valued, we felt safe to ask questions and speak up, so we created innovative solutions.

I loved working in that team. It was the most included and respected I've ever felt at work because I felt safe to make really, really big calls—like moving the patrol, changing our route, saying what I thought the enemy would do—even when they were in contradiction to other more senior teammate's views.

Our Dauntless leader gave me a voice and helped me to make my voice equal, and he did this for many other soldiers on the patrol, too.

I made calls in Afghanistan which were controversial and divergent. I made calls which stopped us driving over IEDs and calls which allowed us to neutralise the enemy before they could attack us. I made calls which were right and calls which were wrong, but even when I made mistakes—and mistakes in that environment have an enormous downside—the leaders around me continued to back me. They put themselves in our shoes and stood beside us. They never left us to take the heat on our own. Dauntless.

My EW colleague and I made critical decisions every day and when on patrol we made them dozens of times every day. Decisions which ensured the physical safety of my colleagues, because I felt psychologically safe to speak up.

I felt safe to speak and I spoke whenever I thought I could offer insight. When I left the Defence Force less than a year after I returned from Afghanistan, I knew that at twenty-eight I'd probably already done the most important work of my life.

This culture where people feel willing to take risks helps individuals grow and develop. Google found many upsides for teams which were strong in psychological safety including improved retention, higher revenue and teams being rated two times as effective as their peers by executives. People who feel safe are better at their jobs because they don't waste time pretending to be something they're not or feeling too nervous to speak up.

SAFETY
[THE APPLICATION]

You can build this type of psychological safety in your team in all sorts of ways, but it doesn't happen by chance. It will take deliberate action to create an environment where your people feel safe to be themselves.

How people communicate

Face to face confrontation can make people uneasy. One organisation I worked with knew they had problems with their customer experience, but when they asked their teams about it, no-one wanted to go into detail.

Team members were worried about complaining, thinking they would get their boss in trouble. The culture there was 'make sure the big boss thinks everything is going great'. No matter how hard the senior leadership tried to get to the truth, people weren't sharing.

Along with a raft of other initiatives to change this culture, one of the tactics they used to get the conversation started was by creating a new communication channel. The leadership team needed to find a way to make people feel comfortable about expressing themselves and looked for an alternative to help people feel safer about speaking up.

A quarterly online chat session was designated a deliberate space for airing challenges and concerns with how the organisation helped customers.

After establishing some ground rules about respectful disagreement at the start, they used this online chat session as a channel to help people raise concerns about their current practices.

The method—an online forum—provided a space more conducive for candour. It was a new forum with the purpose of addressing problems.

The chat room wasn't anonymous—people still had their names visible—but the perceived safety of an online forum helped participants to air issues and float new ideas. The method was the key to opening up a broader discussion.

Off the back of these forums, the organisation redesigned their end-to-end customer experience and improved customer service scores by twenty-one per cent.

Feeling safe to be yourself

Another team I worked in used the first ten minutes of every Monday meeting to ask people about their weekend.

In the meeting time.

On the agenda.

Not as a watercooler chat before we started.

When I first joined the team, I thought my leader was mad. I rolled my eyes 'Who cares about your weekend, let's get on with the job,' I thought. But my leader persisted and with her persistence I soon learnt about my colleagues' partners, children and interests. She drew out our 'real lives' with this agenda item and soon I knew whose footy team had won or lost, who spent their weekends at the art gallery and who was caring for a sick family member.

I learnt what was important to my teammates because the things which were important to them kept coming up every

Monday. This opened a window to a much deeper connection with them. Soon people were sharing their complicated family structures, their health issues, and their children's (or pet's) achievements and struggles. We knew each other. Not in the superficial way we know many of our teammates and not in the way some leaders try to force their team together in team building events or after work drinks. We genuinely knew and cared about each other's lives.

With this deeper connection came more trust, and when I trusted my teammates, I was more willing to share my mistakes, doubts, and concerns. I sought their advice on my projects and when I couldn't find a way forward, I would call them to talk through options. I didn't worry about what they thought I did or didn't know. I wasn't concerned if, by asking questions, I was exposing gaps in my knowledge.

I felt safe they would support or help me and not use the knowledge to expose or undermine me. I felt comfortable with exposing all of myself to them—strengths, weaknesses, nuances, sexuality, family, health and all.

> Dauntless leaders want their team to bring their whole self to work.

With this open frame of mind, we built a large number of ground-breaking, unique and innovative programs that year. We felt safe to share our weaknesses and together we worked to solve problems for each other and openly discuss how to take initiatives forward.

ACTIONS—PSYCHOLOGICAL SAFETY

REFLECT

Spend some time today observing:

- How willing is your team to challenge the voice of the leader or dominant personalities?

- How quickly does your team look to form a consensus as opposed to remaining in conflict or exploration?

- How much of themselves are your teammates willing to share?

GET CURIOUS

Ask three team members about psychological safety.

- What do they think it is?

- How do they describe it?

- What examples can they give of how a leader has encouraged psychological safety in a team they have worked in?

TAKE ACTION

Introduce the concept of psychological safety in your next team meeting. What does it mean to your team and how can you build it further?

Who in your team has great ideas? Do they get invited to the right meetings, and the high-level meetings?

- Invite them—tell your boss you're doing it as a shadowing exercise.

- Make sure they speak in the meeting—follow up with them to find out what their thoughts were, and any gaps they think the 'higher up' leaders are missing.

Review your decisions after trying this for a week. Which decisions could your team have made without you? Where could you have sought the feedback of others? How deeply is your team involved in making decisions about their work? What communication methods could you employ to foster a greater sense of safety and to help people feel more confident to speak up and challenge ideas?

HONESTY
[THE STORY]

Dauntless leaders tell the truth; ugly, direct, uncomfortable or confronting.

"Rach," says Jonno as he munches on a sausage roll after our morning physical training session. He gestures with his thumb over his shoulder towards the office. "Turner wants to see you."

I wipe my hands on my cam pants and put my shirt on before I head up to the demountable building. It's fine to work out the back in just a t-shirt, but you've got to be dressed correctly if you go inside the building. I rearrange my pants at the same time, blousing them in the right spot on my boots.

"Hey, Rach," says Turner as I sit. "I want to talk to you about how you've been doing since you joined the team."

I'm less than a month out from my initial employment training and have been assigned to my first team. I've just turned nineteen and I am green as hell at my job. I know there's been some tension in the team and yesterday one of my teammates had bellowed at me for doing something the wrong way again, but I'm young and arrogant, so I'm expecting compliments.

Turner looks me dead in the eye before she begins. "You see, Rach, what the team is telling me, and what I've observed, is you are being a dickhead."

Sooooo ... not compliments then? I shift uncomfortably in my seat as she continues.

"Here's some things you are doing which are making you a dickhead (she gives me some examples) and here's how it's affecting your teammates (with more examples)."

I wonder how I should respond. I know I can be annoying, but this seems like rough feedback. As I take a breath to start defending my actions, she continues.

"Being a dickhead is not helping our team and while I know you are not actually a dickhead, the way you are acting makes it seem like you are."

She pauses for just a beat before she says, "I want to help you with that."

HONESTY
[THE LESSON]

It seems pretty harsh and you might need to change some of the language choices in a corporate context, but this conversation was not just the kick up the arse I needed but the first time I appreciated how much having a leader who cared about me mattered.

Turner didn't pull me aside because she wanted me to feel bad. She hit me with this brutal feedback because she wanted me to succeed.

She'd spoken with me twice already about my performance before this meeting. She'd offered praise where I'd done well and constructive criticism where I needed to improve. In those sessions she'd shared her own experiences at the regiment and gave me some advice about some decisions she wished she'd made differently early in her career.

I wasn't taking the hint. She could have just left me to keep alienating my teammates and making poor decisions. If she didn't care if I succeeded or failed, she would have. But she did care, so she had the tough conversation with me. Calling someone out like this is uncomfortable. Even in an environment like the military where being direct is valued.

> Dauntless leaders care enough about your success to say things they know you will find tough to hear.

I'd known things weren't going well, but I wasn't sure what to do about it. I'd stubbornly refused to ask for feedback or to ask my teammates for help, preferring to pretend I knew what I was doing. That approach was going badly.

With this conversation Turner provided me a bridge between 'I feel like things are not going well but I don't know what to do about it' and 'Yay—I can talk about this and figure out how I can do better at this job'.

What is valuable isn't just direct communication—after all, anyone can call you a dickhead—what makes the difference is when the directness comes from a place of care. Calling it exactly as it was helped me because the leader had a genuine interest in me.

Direct communication is not a licence to be a jerk.

When it comes from a place of care, direct communication helps you build a culture of truth, trust and growth. It's not just me who thinks direct and genuine conversations are a game changer.

Kim Scott, former Google director, Apple University faculty member and acclaimed coach for companies like Twitter and Shyp, has made a career out of helping people to love their jobs in bullshit-free zones. She calls this ability to tell the hard truth with soft intentions 'Radical Candour'. She speaks about Radical Candour being the zone where you can deliver the kind of direct feedback I enjoyed in the military and have often found missing in corporate life.

Radical Candour is where the care for your people meets your willingness to challenge them and face into conflict.

She talks about receiving some Radical Candour herself after a presentation to the Google founders and CEO about the AdSense business.

I walked in feeling a little nervous, but happily the business was on fire. When we told Larry, Sergey and Eric how many publishers we had added over the previous months, Eric almost fell off his chair and asked what resources they could give us to help continue this amazing success. So ... I sort of felt like the meeting went okay.

But after the meeting, Scott's boss, Sheryl Sandberg, suggested they take a walk together.

She talked about the things she'd liked about the presentation and how impressed she was with the success the team was having—yet Scott could feel a 'but' coming. Finally she said, "But you said um a lot." And I thought, 'Oh, no big deal. I know, I do that. But who cared if I said um when I had the tiger by the tail?'

Sandberg pushed forward, asking whether Scott's ums were the result of nervousness. She even suggested that Google could hire a speaking coach to help. Still, Scott brushed off the concern; it didn't seem like an important issue.

Eventually, Sheryl said, "You know, Kim, I can tell I'm not really getting through to you. I'm going to have to be clearer here. When you say um every third word, it makes you sound stupid."

Being direct can feel unkind, but when it comes from a place of genuine care for the individual and their development your candid, thoughtful feedback is much more helpful than ignoring their weaknesses or problems. It shows you are committed to helping them grow.

Retaining the direct communication style I learnt in the military, while tempering it with less swearing, was key to building the team culture I wanted to work in when I moved into corporate roles. A place where we valued being open and honest. Where my team and I didn't pretend things were fine or avoid tough conversations. It didn't always make me popular with senior management but never backing away from the truth let me work with a team who trusted me to not bullshit them.

It seems redundant to say so, but honesty matters.

When Turner spoke candidly to me and my teammates about our strengths and weaknesses, we as individuals and a group were better at calling out things which were working and things which were not. Same with the combat team in Afghanistan; highlighting failure, weaknesses or areas for improvement was seen and accepted as one of the ways we were building a better team, not taken personally or seen as cutting others down.

HONESTY
[THE APPLICATION]

I worked in the corporate world for more than a decade after I left the Army.

Those closest to me (both my military and civilian mates) were surprised I survived for so long without swearing the house down.

About once every six months or so, an old army mate or a civilian colleague who I worked with in those early years of transition would send me something like this:

> It has been brought to the CEO's attention some individuals throughout the organisation have been using foul language during the course of normal conversation with their colleagues.
>
> Due to complaints received from some employees, this type of language will no longer be tolerated.
>
> We do, however, realise the critical importance of being able to accurately express your feelings when communicating with colleagues.
>
> Therefore, a list of 12 New and Innovative "TRY SAYING" phrases have been provided so a proper exchange of ideas and information can continue in an effective manner.

Instead Of: You don't have a fucking clue, do you?
Try Saying: I think you could do with more training.

Instead Of: She's a fucking power-crazy bitch.
Try Saying: She's an aggressive go-getter.

Instead Of: He's got his head up his fucking arse.
Try Saying: He's not familiar with the issues.

Instead of: And when the fuck do you expect me to do this?
Try Saying: Perhaps I can work late.

Instead Of: Fuck off, arsehole.
Try Saying: I'm certain that isn't feasible.

Instead Of: Well, fuck me.
Try Saying: Really?

Instead Of: Tell someone who gives a fuck.
Try Saying: Perhaps you should check with ...

Instead Of: Not my fucking problem.
Try Saying: I wasn't involved in the project.

Instead Of: What the fuck?
Try Saying: That's interesting.

Instead Of: No fucking chance, mate.
Try Saying: I'm not sure if this can be implemented within the given timescale.

Instead Of: Why the fuck didn't you tell me that yesterday?
Try Saying: It will be tight, but I'll try to schedule it in.

Instead Of: Oi, fuck face.
Try Saying: Excuse me, sir?

I did need to back off on the swearing, but I found I got huge benefits when I found a way to keep the candour and get the best out of my teams.

〜
〜

In the first few months in my new role as Branch Manager, I found myself growing increasingly frustrated with one of my team members.

She was a teller and the youngest member of the team. She was always last in the door in the mornings, scooting in right on start time. She took long lunch breaks and completed her own tasks but nothing more. She made mistakes with her cash drawer and called in sick every month without a reason.

We'd spoken about each of these issues and each time she was defensive and argumentative about the feedback. Little had changed over our time together in terms of her behaviour and our relationship was going from bad to worse.

We met again, the weeks of angst between us simmering.

In frustration, during our tense conversation about her work I stopped worrying about the 'right way' I'd been told to ask questions now I was a civilian and just asked her exactly what was on my mind.

"What is it you want to achieve here?"

She stared at me in sullen silence.

"I'm serious," I said. "We're both saying words but don't think we are really hearing what each other is saying. You come here every day to work with people you clearly don't like in a job you are not succeeding at. Why do you do that?"

She reeled like I'd slapped her in the face.

"What do you mean I'm not succeeding?"

"Your cash draw is always out of balance and you rush to get things done which leads to lots of mistakes. Your teammates are frustrated at your 'bare minimum' attitude. You and I talked about this yesterday just before you left for the day."

"My teammates don't like me?"

"We've talked about this in our last meeting. And the one before that."

"No, we haven't," she insisted.

I took a deep breath, feeling like a broken record.

"It's not so much they don't like you but when you can't balance everyone has to stay late to figure out the problem. When you arrive at the last minute, we can't get set up for the day because you have the main key. Your behaviour makes them think you don't care about them and the things they need to do both here at work and at home."

She changed from defensive and angry to horrified. I'd not been this direct before and the impact of being totally open about how her behaviour was affecting her teammates cut right to her core.

She was upset, but slowly she opened up.

She talked about juggling her work and her studies.

She talked about her home life and the things making her late.

She talked about how much she loved her teammates and her dismay at letting them down.

She talked about her career goals after she finished her degree.

It turned out she was taking sick days to complete her studies. Her previous manager didn't want to do the paperwork to get her study leave approved, so told her to just take sick days.

She confessed she'd never spent any time thinking about what she could achieve at the bank because she'd just considered it as a way to earn money while studying.

> No-one had ever asked her what she wanted to achieve

Her previous leaders had just assumed she was here for the short term because she was studying, so they didn't give her any of the key clients to look after or any of the longer-term jobs. She had no additional training and hadn't been considered for promotion. People had made a bunch of assumptions about her and, based on those assumptions, she had decided there were no opportunities there for her either.

Although the conversation went better than any of the others we'd had before, there was still plenty of tension we needed to work through together from our earlier miscommunications.

She later told me she walked away from our conversation thinking, *I'll show you.*

And show me she did.

By the time I took my first promotion at the bank a few years down the track, she had been promoted twice, won customer service awards, won team awards and was appointed as Bank Manager to replace me.

I loved seeing it.

It turns out my frustrated question was a good question. It was one I was able to use various iterations of throughout the rest of my career to open up an honest dialogue with people on my team. I wasn't looking for a prescribed answer, I just focused on being genuinely interested in what people wanted to get out of their work.

"What is it you want to achieve here?"

Some wanted to get closer to their customers. And that was cool.

Some hoped to move into different departments. And that was cool.

For some, the work complemented their studies, giving them practical experience. And that was cool.

Some wanted to be the best at their job, some were there just to put food on the table. Some wanted instant promotions, others had done their role for twenty years and never wanted to change. And all of that was cool.

One girl told me she wanted to become a florist, so we sent her on all the business banking training. She learnt about profit/loss, cash flow management, credit systems, and payments solutions. She looked after all our local business customers, asking them about their challenges and what they had learnt working for themselves. She was genuinely interested in their businesses and was learning all about how to run her own business at the same time.

By the time she stepped out into her own florist shop she had built a business plan which covered every detail of her first three years.

And that was cool. I loved seeing it.

My team's honesty allowed me to provide them with the right opportunities to pursue their goals and to plan for the future as they moved through the organisation. Without honesty they miss out, and I'm left with unexpected gaps in the organisation.

When you can get to the heart of why people are doing what they are doing—what they hope to achieve (at work and in life) and learn not to take it personally if their path is with you or not—then you open up a whole other level of genuine conversation.

Radical Candour is not just for leaders. When the whole team can do it too, it's a key building block for trust.

ACTIONS—HONESTY

REFLECT

What tough conversations have you been avoiding?

Is there someone in your team you believe has potential but isn't achieving it who could use some Radical Candour?

GET CURIOUS

Ask three team members:

- I want you to tell me the raw, honest, unfiltered truth—how do you think we performed as a team during our last project?

- What stops you from providing unfiltered feedback at work?

- Would you value honest feedback about your performance, even if it was tough to hear?

TAKE ACTION

In your next one-on-one ask everyone, "What do you want to achieve here?"

Work with them to create a plan of how to get there. Come to an agreement on how honest you'd like to be with each other about how the plan, and their performance, is going.

TRUST
[THE STORY]

Dauntless leaders communicate with clarity and trust their people to do their jobs well. They give their power away.

Volleys ring out across the dusty, grey-brown valley and echo into the empty space between the Forward Operating Base (FOB) and mountains of the Hindu Kush.

A Dutch patrol is firing at the eclectic targets which make up the rifle range on the base the Australian and Dutch forces share—Camp Holland—in preparation for tomorrow's patrol.

A bonnetless, pockmarked station wagon sits on bent and crumpled rims with two giant, fluro-pink numbers—a seven and zero—spray painted on its flank to indicate its range of 70 metres. A 70s era Russian tank sits hulking amongst the 300 metre targets.

Yellow corflute panels printed with body-shaped silhouettes peek out from behind structures. Some are zip-tied to star pickets pounded into the open ground. They look like spotlit actors on a stage, almost naked in their vulnerability, and it feels mean to pick them off as they stand so exposed in the open. Earlier, when I'd calibrated my weapon at the range, I went for the hiding silhouettes, though I knew none of the plastic images were judging if I was being fair or not.

In sneakers, shorts and a brown t-shirt, I hold my rifle in one hand low at my right thigh as I jog past the Dutch contingent practicing at the range.

Exercise is a great way to burn through down-time hours on operations and I pass half a dozen other soldiers also out running on the ring road and dressed in the same strange getup of shorts and weapons. If I push hard enough on the track, I can sometimes distract myself from thinking about how much I miss my family. My son will be starting school this year and I already know I won't be home to see him off on his first day of school.

The gritty five hundred metre long circuit parallels the bomb-proof bunker walls surrounding the base and provides the only outdoor running option in on the base. I enjoy the steady crunching rhythm of my feet as I trundle through the perpetually beige landscape.

A Bushmaster rumbles past, then its bigger heavily armoured brother, an ASLAV, with its eight tightly packed wheels churning up the dust. I hack and cough in their wake. The heat of summer lingers through the early autumn days and sunrays reveal swirling dust so fine and light it fills the air. It is so dry no matter how much water I drink I always feel like I've been chewing on cotton balls.

I hear a burst of rounds behind me and spend the next few strides thinking about how different controlled single shots sound from machine gun fire. I wonder if we'll hear either tomorrow. It will be the first patrol for 1st Reconstruction Task Force soldiers. Special Forces will come with us and provide guidance.

I push up the rise and breathe heavily at the crest of the hill. I veer off the track to lean against the dirt-filled barrier while I suck in noisy gasps of air. Just a few steps behind me my boss pulls in too. Benno looks fresh-faced and is barely panting. I hate him a bit for how easy he finds running.

Together we look out over the valley towards the village I will patrol to tomorrow.

"You are genuinely terrible at running," Benno says as he casually throws his leg onto the barrier and stretches out his hamstrings.

I'm gulping in breaths I am half-collapsed over the bunker wall. My maximum effort response is to raise an eyebrow.

"You work at it like a determined bastard though," he offers. "And though you lack running skills you are excellent at helping us avoid the horse-meat salami lunch option."

The menu at the Dutch mess has taken some adjusting to, and we'd all eaten horse-meat salami for days before I'd worked out what the Dutch on the packet meant. In the end we decided all fresh ration options were better than ration packs, even if they were strange fresh rations.

"Also the first EW pick for the patrol tomorrow," Benno says.

I nod and gulp at the water he offers me from his canteen. The hot, dry wind whips at our faces at this exposed edge of the base and wicks away my sweat instantly. We've only been in-country for a week, but it feels like the permanent weather prediction here is 'imminent sandstorm'.

"How are you feeling?"

I'm no longer the eager nineteen-year-old venturing into risky situations with an excited attitude. It's been seven years since I deployed into East Timor. Since then I've had a baby, been posted to Melbourne and Canberra, spent two years working at the Defence Signals Directorate as an analyst, completed all my promotion courses and am now a team leader myself.

"I want to do a good job," I say.

I hand his canteen back. "I wish we were more prepared; I wish we knew more about what and who to listen for."

We stand beside each other in silence for a few moments. It is always the challenge of being on the first deployment into a new country. You've got a very scant signals intelligence picture when you arrive because you are the team who builds the signals intelligence picture. I was among the first into East Timor too, so I know there is little we can do other than to start.

The sun is setting behind the mountains and the sharply ridged shape of their peaks appears in a block of shadow creeping across the valley floor. Even before it is consumed in the gloom, the vista towards the village is Star Wars-esque—desolate, grey and foreboding.

"I don't want to make a mistake," I say as I chew at my dry and chapped lips.

We turn back to the road and start jogging together down the hill.

The day we arrived at Camp Holland I'd run into an old EW colleague. He'd been attached to Special Forces and in-country for months before the arrival of the Australian Regular Army deployment—the 1st Reconstruction Task Force that I was part of. In covert conversations he'd given me as much information about what to expect as he could. I knew there was a real chance of volatile situations turning into skirmishes, with indirect fire attacks on the base and IED threats ever present too. Based on his experiences, I also knew if we could intercept the right information we could make a real difference to the safety of our team.

It is easier to be honest with Benno now we are shoulder to shoulder.

"I don't want anyone to die because I make the wrong call."

Benno served for twenty years in the UK before transferring to the Australian Army. He has retained his rank and brings with him a wealth of experience including his time as part of the United Nations Protection Forces during the Yugoslav Wars. He has been there, done that.

"All you can do is make decisions based on what you have," he says as the ground flattens out and we loop back towards the Australian part of the camp.

"You're never going to have the luxury of waiting until you are certain and you're not going to have the time to check with us back at the base. Make the calls while you are out there with the patrol. Back yourself."

Light is fading as the Australian flag comes into view around the next bend. I hear a whistle in the sky and it takes a second to register it as an incoming rocket. We dive to the ground against the barrier wall as the thunder of the vicious impact booms, thwacks and echoes to the east of the base.

The sirens set high on every corner of the walls surrounding us begin to wail and we run towards the nearest shelter. The shelter is a blue shipping container half dug into the ground with sandbags stacked high up the sides and onto the roof. The base is instantly full of soldiers in various states of dress streaming towards their allocated bunker.

I see a mate who's been sleeping after his night shift running towards us with wild bed hair, in thongs, boxer shorts, and body armour. He scoots in the door and, knowing he looks ridiculous in his eclectic mix of civilian clothing and military weaponry, he takes the time to turn back and gyrate his crotch at us with a big grin on his face before he disappears inside the shelter, his rifle clutched in one hand and a Kevlar helmet in the other. I bark out a laugh. There is always time for comedy.

As we reach the door I see a dozen soldiers already hunkered down against the walls, behind me a few others straggle in from the plywood shelters we sleep in. The huts are much more comfortable and sturdy than being under canvas, but they offer little protection against rocket strikes.

At one of the huts a flimsy wooden door is flung open and smacks shut on its return spring as a soldier launches himself from the top step, already running as he hits the ground. The sound is the same one the old fly-screen door at my grandparent's house used to make; cracking like a whip when it slaps the frame of the door.

The Dutch Quick Reaction Force (QRF) vehicles start up and race out of the base. Tyres spin and back ends sway as they storm out the gate on dirt roads and across the valley in the direction the rocket was launched from.

Someone radio checks into headquarters from our bunker, counting off our soldiers and accounting for those on shift or ring-ins like Benno and I who've bundled into a different shelter.

After twenty minutes with no further action we are stood down and everyone heads back to what they were doing before the interruption. It's surprising how 'normal' this all feels even after only a few days in Tarin Kowt.

≈

The next morning we prepare our vehicles before the patrol.

Special Forces have been in and out of the country for years but this is the first regular Army deployment to Afghanistan— to Tarin Kowt in the south of the country—and our first patrol.

There's about a hundred of us who are preparing to roll out of the gates of the compound for the first time. The bunker walls at the gate are about three stories high; tonne upon tonne of rammed earth packed into big hessian-lined wire cages. The hessian adds to the overall beigeness of the landscape. About a dozen vehicles are lined up in a big open area inside the huge walls.

This staging area is about as big as a football field, but it looks nothing like a football field. There's no green, no grass. It's dirty and dusty and dry. I can feel the grit and dust in my teeth, in my hair, and in the sweat that runs down my back underneath my body armour.

I check my rifle. Weapons need constant attention in the dirty, hot environment, and it's hard to find the balance between using enough oil to lubricate the working parts, but not so much you turn the whole thing into mud.

We're waiting for the order to move.

I can see the nerves pulsing through everyone around me and my own heart is definitely racing.

Soldiers in full body armour move around restlessly. Some wander between groups, others stay where they are but shift from foot to foot. Some jabber to their teammates, their jaws working off their nervous energy. Others retreat into their own thoughts.

There's a tonne of bravado; back-slapping, high-fiving, trash-talking, locker room behaviour. About a thousand cigarettes are smoked and shared.

I'm thinking about the job I need to do.

And I'm questioning myself.

You don't really know how you will go when it all goes down. You hope you will be brave; but you don't know until you're there, until you know.

That's what I think lots of the nerves are about, not just mine.

We've completed extensive orders earlier but given this is our first time out the gate the commander gathers us together again before we leave.

"Okay team, simple mission today for the first run out. We secure the Tarin Kowt hospital compound, meet with senior medical staff, then come home. We're not fixing anything today, just gathering information and getting back safe."

He makes eye contact with soldiers among the group.

"You each know the mission, you each know how your role contributes to the mission."

He pauses. My mind races to our EW gear, our plan for search, what we need to do if we find anything which threatens the security of the patrol.

"Five minutes notice to move," he says. "See you all back here when we're done."

TRUST
[THE LESSON]

It was a real challenge for me to choose the right story about trust.

Almost everything the military does builds trust. I feel almost every story in this book showcases how critical trust is in military teams and how much work goes into building trust in every team. From how teams are built and led to how soldiers are trained. How we talk and interact with each other and how empowerment is not a buzz word but a critical part of even the most junior soldier's everyday life.

Even things like people's rank and corps badges tell us a whole heap about someone we are interacting with—like how long they've been in, what experiences they've probably had, what expertise we should expect them to have—without ever having met them before. All this builds trust. I trust them because I extrapolate the experiences I've had with other people in that corps or with that rank onto the new person I'm working with. I transfer trust I've built with others onto this new teammate who I know from their badges shares similarities with my previous trusted teammate.

Then there are the shared experiences, the shared hardships, the wins which are hard won and the losses which you all feel for decades afterwards.

On top of all of that are the deliberate actions you take together to build trust.

That's what I looked for in the right story for this chapter. I wanted to look beyond the inherent trust military members share and find the deliberate actions Dauntless leaders take to build trust. I believe there are three things the military does exceptionally well to actively shape a culture of trust in teams:

1. A concept called Commander's Intent

2. Empowerment at every level

3. Emotional Intelligence

Commander's Intent

Commander's Intent provides teams with a clear description of the desired end state.

It's what the battlefield should look like at the end of the mission. It is what success looks like. For us on that patrol: *Secure the Tarin Kowt hospital compound, meet with senior medical staff, then come home.* It works because it focuses on **what** needs to be achieved, without prescribing **how**.

I bet your organisation already has a mission. Maybe it's called a vision. Maybe you have both.

It'll be at the top of your 'About Us' page.

I guarantee it's repeated ad nauseum by senior managers.

I envisage it written in calligraphy on a prominent wall in the head office.

And I'll assume it's full of vague, broad words and sounds good, but no-one really knows what they are supposed to do with it, right?

That's what most of the ones I've seen are like.

Maybe you know your organisation's mission by heart? Whether you do or don't (or your team members do or don't), reciting the company mission or vision is not enough, and in fact, is not the point.

What Dauntless leaders get right is they focus on people **understanding** what the mission is. They worry about getting people onboard with the big goal, and then get out of their way when it comes to execution.

Dauntless leaders create an organisational shorthand in their teams to help people understand the mission. It is the *meaning* behind the words they focus on, reinforcing it with every interaction. Who Dauntless leaders compliment, how Dauntless leaders praise, what Dauntless leaders reward all reinforces the meaning behind the words they say matter in their team. If your mission has twenty words and lots of clarifying words, it's a sure indicator you don't have an organisational shorthand.

To use Commander's Intent effectively you and your team must first spend time to create a common understanding of the meaning behind the words—your organisational shorthand—because when people know what they are *really* there to do, they can keep making decisions which push them towards the common goal.

> Knowing exactly and explicitly the meaning behind the mission statement helps you to create an organisational shorthand.

We'd done lots of work prior to our patrol so we all understood exactly what 'secure' meant when we needed to secure the hospital compound.

Most of the corporate and business missions I see are some variation on 'Helping Customers' and this is a great mission. It's full of purpose and many people find meaning in work and a connection to their personal values in providing service to others.

However, you need to scratch beneath the surface of each word in the mission statement to get a true *understanding* of it, does 'helping' mean:

- Cordially greeting every customer?

- Doing it for people, or teaching them how to do it for themselves?

- Being fast?

- Getting it perfect?

- Demonstrating care and empathy?

- Improving productivity?

- Recommending services and products?

- Supporting the client to make their own choices?

- Facilitating?

- Coaching?

- Aiding?

- Or does 'helping' really just mean 'sales'?

It might mean all these things. Whatever the meaning of 'helping' is in your organisation, exploring how your team members understand it will tell you a lot about how clear they understand your mission.

Turning a vague mission into Commander's Intent is about digging into the details of what the mission means to your people, your teams and your organisation. These words and meanings become your **organisational shorthand.**

> To communicate your intent with clarity, you must do the work to ensure each member of the team knows the *meaning* behind your mission.

Military leaders trust their teams enough to provide them with the true goal—secure that compound, take that hill, get that info—and why. Because if your team on the ground have to constantly come back for decisions they'll probably get killed before they get an answer. If you trust your team with the end goal and they truly understand what the mission is, they can make decisions on their own.

Empowerment at every level

> Dauntless leaders look for ways to give up their power.

Dauntless leaders know they build the best culture and teams when their time is spent thinking about the big picture and letting their people have the power when it comes to execution. They let local leaders and subject matter experts decide on the details and they do this because they know it makes for better results and it reinforces trust in their teams.

You build trust by giving trust. When you focus on the big goal and let your team execute the way they think is best, you're actively, publicly demonstrating your trust over and over again.

When people build the plan they buy into the plan. Setting a clear goal but letting others decide the details gives them both ownership and accountability. They'll want it to succeed because it is their plan.

You can see this in the combat team commander's approach. He didn't detail every step of what he wanted us each to do but instead he talked about what things should look like when we were done:

Secure the Tarin Kowt hospital compound, meet with senior medical staff, then come home.

You can see it with my boss, Benno. Rather than trying to hold on to power he encouraged me to make my own decisions: *On the ground you'll see and feel what is happening around you; you guys on the patrol are the right people to be making the calls.*

Although he had more experience than me, more expertise and was a more senior leader, he knew making me come back through him for decision making wasn't going to improve our force protection and wouldn't help me grow. He gave his power away. Dauntless leadership.

The military pushes power down the chain because of the number of variables soldiers face during action. You practice scenarios (like we did for the helo insert into Suai) but you can't cover every possible combination of events. Dauntless leaders concentrate their efforts on ensuring everyone knows the end state and then they trust people and empower them to make decisions along the way in a changing environment.

There were lots of variables about how things might go down once we set out on the first patrol in Afghanistan. We didn't know which local leaders would show up, if they would talk to us, if they wanted to talk, how long they would talk for, and about what.

We didn't know what dangers to expect either. If insurgents knew we were coming for the meeting what actions would they take? Would they try to attack the patrol on the way in, or the way out, or even while we were in the compound? Or would they just observe? Even entering and securing the compound would rely on us making decisions on the fly once we arrived, because we hadn't been there before.

Now I'm not saying the military runs around without a plan.

We rehearse scenarios over and over again so we can react to changing and volatile situations with precision, but the purpose behind practice is not to get things perfectly choreographed for every situation. The value of rehearsal is it familiarises people with variables and builds awareness of how we need to communicate changes and to who. Rehearsal is about building our capability to be flexible in our approach when it comes to the real thing.

> Dauntless leaders trust their people to make decisions once the plan is in motion.

In the midst of a firefight I don't have time to get forty-five people sign off on a one-page memo. Changing the font from Arial 11 to Times New Roman 12 isn't going to make the difference between a good decision or a bad one. Especially if that decision is being filtered back up through the chain to be made by someone who is not even there on the ground.

Dauntless leaders do the work to make sure their people understand the mission and understand the role they play in the success of the mission, then let their team get on with doing the doing.

Emotional Intelligence

Notice how much space and silence Benno gave me during our jogging chat.

I can assure you, it wasn't because he was trying to catch his breath! I don't think I've ever seen him look remotely like running was an effort at any time.

Whether he'd label it EQ or not, that is what was happening here. Benno knew me well enough to simply leave space and wait. I'd keep talking through my deflection answers until I got to what I was actually thinking about or worried about. He didn't know me 'long-enough', he knew me 'well-enough'; we'd been working together for less than a month. He'd spent that month getting to know each of us in the EW team—our styles, our behaviours, our personalities—so he could coach and support us to be successful when we were heading out on patrols.

Benno's tactic of space is one which is useful in coaching, supporting and leading many people, not just your 'Rach's' who are verbal thinkers and deflectors. Giving people the time to think deeply about what they are doing or feeling helps them work beyond their automatic responses with what they think you want to hear, and to move towards their true thoughts.

Dauntless leaders are not afraid of silence.

In my conversation with Benno, I first offered up a worry about *doing a good job*. Then I made some excuses about not being prepared before saying *I don't want to make a mistake.*

Finally, I said what was really bothering me *I don't want anyone to die because I make the wrong call.*

I was worried about my ability to make the right decision in a critical situation.

If Benno had addressed any of the other concerns I'd rattled through, we wouldn't have dealt with my real worry before we rolled out the gates because I never would have revealed it.

With Benno's advice I made a lot of decisions. I made the calls. On that first patrol and every time after that I went out the gate. Six months later I received a Commendation for excellence in leadership, analysis and advice under extreme duress in a hostile environment—all because Benno had encouraged me to be confident and to back myself.

> Dauntless leaders know when their people win, they win.

Often we're so enthusiastic about helping or want to share our knowledge and make the journey easier for our people that we offer solutions far too quickly. People don't benefit when they are told what to do. They learn and grow when they find and uncover their own solutions. When we go into solution mode too quickly, we stop people from self-discovering. The other issue with moving to solutions too quickly is we often haven't dug deep enough in the conversation to really find what the root cause is.

When was the last time you let someone sit beyond uncomfortable silence before you filled the space with words?

Great questions work to help people uncover their real concerns too. Not questions which try to guide your teammate towards the answer you want, but curious questions which you genuinely don't know the answer to. This is where you can help people to get to their own truth and build trust.

> Dauntless leaders are genuinely curious about their people.

Dauntless leaders are the leaders who want to know what makes their teammates tick, what they care about, and what they want to achieve. You can feel it when your leader cares about you as an individual, and if you lead a team, your team can feel if your care for them is genuine.

I believe this is the key to unlocking your EQ capabilities—curiosity.

If Emotional Intelligence is recognising, understanding, and managing our own emotions along with recognising, understanding, and influencing the emotions of others, being genuinely curious about other people will take you a long way toward achieving this.

TRUST
[THE APPLICATION]

I've seen some cracking examples of trust outside of the military too.

One Dauntless leader I know designed a plan for their teammate to work offshore as a digital nomad in an organisation where most people struggled to get permission to work from home one day a month. Both leader and teammate focused on how the tasks could be achieved (not why they might not be able to be achieved) and forced their way through the red tape with a 'just because it hasn't been done before doesn't mean it can't be done' attitude.

This Dauntless leader took the view if their teammate could maintain their outcomes and customer satisfaction scores it didn't matter where they did it from. Together, they were open and honest about what worked and what didn't. When the digital nomad walked off the plane and found herself on an island with no wifi, she immediately booked the next flight out of town to somewhere she knew she could connect. The respect and trust between the two of them was mutual. They both knew the upside not just for themselves, but for the whole organisation, if they could prove it could work. It did work, and remote working became part of the everyday culture in their organisation.

Or the time when one of my peers felt safe enough at work to point out the flaws with one of our project roll-outs just days before it was due to launch. Although the team had reviewed the concept together and agreed on an approach, she didn't feel comfortable with the end product. In the final review before launch she stepped up to say 'now I see the completed product I worry this approach feels condescending and potentially offensive to some of our people.'

We'd come so far, and were almost completed, but she spoke up anyway. The work our leader had done up in the months before this to encourage productive and respectful challenging among our group saw our usually introverted team member voice her concern at a time which was pressurised and public. We stopped, reviewed, and changed the approach. Our team member was right, and her voice ensured we avoided a huge embarrassment.

Another leader told her top performer he had so much more potential than he could achieve in his current role and promptly sponsored him into a job where he could stretch despite the gaping hole it left in her team.

The leader who trusted an intern to lead his team. The leader who promoted someone who didn't look like a traditional fit for the role. The leader who told his team he'd lost his passion for the business and was stepping down.

Everyone must **feel trusted** and **trust others** for the environment to be one where trust thrives.

I'd found a leader cultivating trust early in my corporate career too. I'd grown bored with branch life in Toowoomba and wanted to spread my wings. My boss didn't have the right next role for me, but he introduced me to someone in the business who did.

> The multi-story head office building was smack bang in the middle of the CBD. I'd only ever been to the fifth floor for training so getting off on the third, knocking on a random door and meeting a guy I'd never seen before had me hoping I'd navigated to the right place.

"Tell me a bit about yourself," he asked after I sat down.

I launched into my story of the Army, my leadership experiences in the military, then detailed my years at the Toowoomba Branch.

Within the first ten minutes of our conversation he waved his hand at me.

"Look mate, I've heard enough. You'd be great—just what we need—so let's just cut to the chase. I need a branch manager in Cairns. It sucks up there at the moment and people are unhappy so I need someone who will tell me what the hell is going on and what we need to do to fix it. I also want someone who'll go to all the meetings I don't want to go to."

I laughed out loud. The refreshing honesty of it all was an absolute delight.

I was still a relatively junior bank manager, and although our branch was performing well, what he was talking about was a big step up. Those meetings he didn't want to go to were with the Managing Director for the State and he was talking about giving me an opportunity to position myself as the next regional manager.

Even though I barely knew the bloke, I could feel it might be a great fit. He'd already given me his commander's intent—the things he wanted me to deliver on—and he'd clearly indicated he trusted me to do the job.

"Why are you looking for someone?" I asked.

"Last leader there couldn't get the team behind him. He was only in it for himself."

"Well I'm about the team."

"I know. I've already asked a few people about you—including your staff—so go home, talk to your family and if you're interested, tell me what you want to get paid and we'll sort it out."

I was gobsmacked.

If I didn't know him, a senior leader in the organisation, then how did he know me? Or have the time to speak to my team?

Before I could answer he said, "I know it's the middle of the year and probably not a good time to move schools. Your son's in primary school, right? Just let me know what your family needs to be able to make it work if you want in."

I'd found another Dauntless leader who was bold, honest and willing to back their people.

He was clear about what he wanted me to achieve—Commander's Intent.

He wanted to give away his power—Empowering others.

He'd already tried to find out what was important to me—Emotional Intelligence.

With direct communication (I was never left wondering if I was doing well or poorly), masses of freedom for me to run my own branch how I wanted to, and a genuine interest in who I was and what was important to me, plus an intent to help me achieve my goals—I bloody loved working for that bloke.

He also had a knack for gathering a broad range of diverse leaders to work with and for him. Men and women who were kind and encouraging, in counterpoint to his gruff directness. Others who led his largest teams, who challenged him, argued with him, and helped the team drive for ever greater outcomes. People from outside the industry and from different age groups, backgrounds, and geographies. He embraced remote working without a second thought. If you were the right person for the role and could deliver the outcomes, then how and where you did the work were all easily navigated without fuss.

My new teammates all quickly embraced me and found ways to include me and welcome me into their team. It was just how things were done, if you are on our team you're one of us now. There was no 'one size fits all' leader or team member. Our Dauntless leader trusted and believed in people first, then worried about any issues later *if* they occurred.

We don't do that enough. We assume people are going to do the wrong thing, so we create pages and pages of policies to prevent people 'abusing the system' when in reality most people do the right thing. What we do when we create all these hoops for people to jump through is slow them down, make it hard for them to get their job done and show them we don't *really* trust them.

My potential new leader's question about our family situation sent me home with a pile of questions and possibilities on which I mused during my two-hour drive back to Toowoomba.

When I got home, we did what we always do in our family, bring it to the table.

At our house there are a few rules:

1. **Everyone must agree with me when I say I am tall.** I'm 165 cm tall but in my mind I'm a giant, hulking dominating force of nature. So get on board and when I talk about my tall-girl problems, nod and smile and say 'yes Rach, it must be hard to be super tall' (even if you are taller than me, which everyone in my family is).

2. **The default option is 'yes'.** We noticed so many other families, leaders and people with a default position of 'no' we decided we didn't want to roll like that. 'Should I try to do a backflip on the wakeboard?'—yes. 'Wanna go kick the footy down the park on Saturday?'—yes. 'Ice-cream for dessert?'—yes (if you eat your dinner). You might think these questions came from our son but they're actually all questions I've posed to our family

group. Our default answer is 'yes' and we only say no if there is a *reason* to say no.

3. **Openly declare who has the primary career.** It's not necessarily who earns the most. It might be whose industry has the most interesting opportunities at the time, or who is involved in the most meaningful work. Rather than have people feel undervalued or worry about what they are doing next, we openly declare who has the primary career at any point in our house. Sometimes it's been me. Sometimes it's been Damo. In the last three years of his senior school, it's been Will. It's always up for negotiation but knowing whose career has the most prominence at a given moment in time, helps us to make collective family decisions.

4. **All opportunities must be brought to the table.** Bring 'em. No matter how crazy, challenging or in conflict with the primary career they might be, bring 'em.

So I bought my opportunity to the table.

Despite the fact Damo had the primary career at that point— he'd served 15 years in the ADF and he'd likely have to give up the job he loved—we talked openly and honestly about our fears, the opportunity and where the way forward could lie for our family. We switched who had the primary career, and we moved to Cairns.

I've been lucky to not just have trust in my workplaces but to have a family which supports and trusts me too. If you exclude my rule about 'agreeing I am tall', everything else about our rules at home tie into the lessons I've learned from Dauntless leaders about trust. Open, honest conversations where everyone feels valued, calls it like they see it and is included in making decisions, make a big difference to how your team (like your family team) pulls together.

Trust takes lots of elements to get right and needs your constant attention, but if your team knows where they are going, have the power to make choices about how they get there, and you show you give a shit about them, you're well on your way to a great trust environment.

ACTIONS—TRUST

REFLECT

Consider the three elements of trust:

- Commander's Intent—how clear is your mission? How deeply does your team understand the *meaning* behind the words?

- Empowering others—What decisions could you push down the chain? Who on your team would benefit from the additional responsibility?

- Emotional Intelligence—When was the last time you let silence do the talking for you? How well do you know your people's preferences and drivers?

GET CURIOUS

Choose three teammates who you feel you know least.

Ask them:

- In the best teams you have worked in, what made them great?

- What is the purpose behind our work here? What are we trying to achieve?

- What are the things you feel you should make the decisions on which you currently have to get sign off for?

Record their ideas and reflect on their responses.

TAKE ACTION

Explore your mission/Commander's Intent in your next team meeting. What do the words behind it really mean and are you all *really* singing off the same hymn sheet?

What are three decisions you've made which would be better sitting with your team. Find a process which lets you delegate these decisions going forward.

What don't you know about your team? In your next one-on-one ask each person a question you don't know the answer to and explore the path it takes you down.

REFLECTIONS ON CULTURE

No matter how inspirational your speeches or how compelling your purpose, if the culture is shit, it undermines the lot.

Culture is always in flux with personalities, leadership, and the external environment, all influencing and impacting on the culture of an organisation.

> Dauntless leaders know culture needs their constant attention.

Creating a great culture comes first for all the Dauntless leaders I've worked with. Before they worry about their individual actions, they focus on the right environment for their team to thrive. They focus on the elements of safety, honesty, and trust within their teams, knowing these are essential to the culture of a thriving team. It's similar to Maslow's hierarchy of needs. Without excellent psychological safety, team honesty and a culture of trust, you can't build a team.

> People can only focus on their own development and excellence when they first feel safe to be themselves in an environment of honesty and trust.

Culture is not an activity or an offsite event which can be completed and then marked as 'done'. It will take your constant vigilance and continued effort to nurture and grow the culture which will help your people and your business to thrive.

> Our behaviours shape our culture and our culture shapes how people behave.

Culture and behaviours go hand in hand.

Culture and behaviours form a symbiotic relationship, feeding off each other which can build, becoming either increasingly positive or increasingly negative.

Over and over culture and behaviours reinforce each other—what people do reinforces what is acceptable, and in return, what we say is acceptable is reinforced by what we see others do and how they do it.

There's a parable about five monkeys, a ladder and the social transmission of acceptable behaviour which illustrates this cycle.

Five monkeys are locked in a cage, a banana hung from the ceiling and a ladder placed underneath it.

Anytime a monkey attempts to climb the ladder, researchers spray the climber AND the other four monkeys with ice-cold water.

When a second monkey tries to climb the ladder, the researchers again, spray that monkey AND the other four with ice-cold water.

This is repeated until the monkey's learn—climbing equals cold water for EVERYONE—so no-one climbs the ladder.

The researchers then replace one of the monkeys.

The new monkey attempts to climb the ladder to get to the banana, but the other four monkeys stop her—corralling her away from the ladder and physically restraining her from attempting to climb. If any monkey attempts to climb the ladder, the other monkeys attack it to stop it climbing.

The researchers do not spray the new monkey, or the remaining four. The group enforces the learned behaviour on the newcomer.

Over the next few weeks, new monkeys are introduced into the cage, replacing the first cohort. Each time as one approaches the ladder, the other monkeys restrain it and prevent it from climbing until not a single monkey in the cage has ever been sprayed by water, but none of them attempt to climb the ladder.

No-one has told the monkeys their culture is a non-ladder-climbing culture. However, each new monkey joining the group sees only non-ladder climbing behaviour is acceptable.

Newcomers learn the behaviour and culture of an organisation from the people they work with, even if they don't understand why, or the history of why.

Regardless of how many times you tell your new hires you love people who climb ladders, if all they see are ladder-climbers being suppressed, reprimanded and restrained by their teammates, no-one will climb the ladder no matter what font size you print your ladder-climbing-values on the company wall.

UC Berkeley development psychologist, Alison Gopnik, talks about the importance of environment in her book *The Gardener and the Carpenter* when exploring the relationships between parents and children—and her analogy works for creating culture too.

The parent or leader who views themselves as a carpenter spends their time shaping others; chiselling and sanding away at the 'raw material' to shape them as they see fit.

Those who are gardeners spend time nurturing a healthy environment to help individuals in their care to grow—including tailoring the conditions depending on what species of plant their charge might be. After all, a daisy seedling needs different water, light and care than a towering, established eucalypt.

Carpenters fail to take into account the unique capabilities and character of the individual they are developing—rather looking to wedge and shape them into a preconceived ideal of what a 'great child' or 'great employee' might look like.

Gardeners don't try to change or create the individual. They focus on what helps a particular plant to grow and flower; they nurture the environment around them to ensure optimum growth conditions.

> Culture needs this same 'gardener-like' attention; a constant tending.

Dividing and replanting those which have grown too big, nurturing and supporting those who have just started out.

If you leave it alone and don't tend to it, you'll come back in a years' time to find the voracious fast-growing plants have squeezed out your quiet achievers until they are twiggy and thin.

Your beautiful flowers will now hang their heads, weeds of distrust running rampant over the paths and the harmony of shade and light, damp and dry now just barren dirt. Symbiotic relationships no longer thrive, bullies come to the fore.

No-one wants to work in that garden.

Spend the time to cultivate your culture.

CULTURE—where to now...

REFLECT

Review your answers from the start of the culture section.

- How much trust is shown and given in your team?

- When was the last time someone told you the honest, unfiltered, ugly truth?

- How do you give away power?

GET CURIOUS

Get a reality check on your current culture. Review the team feedback on culture from the first section and ask three more people about team culture:

- What do they believe your team does to build culture?

- What else would help promote trust and honesty in your team?

- When is a time they have held back from being totally honest with a teammate? Why?

- How much power do they have to make decisions? What percentage of their tasks are they in charge of and what percentage do they need sign-off on?

Consider if you were an outsider looking in, what do you notice which aligns or doesn't align between observed culture and your desired culture?

TAKE ACTION

Put your aspirational culture notes to the side and schedule an appointment with yourself to review them.

At the appointment:

- Reflect on the feedback and insights from others and incorporate their ideas into your aspirational culture statements.

- Identify three actions you could take to build the feeling of psychological safety in your team.

- Be honest about how much you trust your team to challenge you and to make their own decisions. If you are holding the reins too much, explore why. Make a plan to delegate decision making this week.

- Consider the gaps between where you want your culture to be and where your culture is now—define one action each month you can take to close those gaps. Decide how you'll measure the success of your actions.

TEAMWORK

Dauntless leaders encourage different thinking and understand inclusion is the key to harnessing diversity of experience and thought.

When you are in the position to hire other people, it's tempting to pick yourself as every member of your team.

Your ego is boosted when everyone thinks the same way as you do and affirms all your decisions are the right decisions. You feel like you are all so in sync but there are significant disadvantages to being a team of clones. You'll only ever get slight increases in the way you achieve. You'll never see massive innovation or harness the giant leaps which diversity of thought can bring to your team. You might relate to each other about the type of schools you all went to or your experiences growing up, but you'll never get a unique perspective on an issue or be able to consider how a broader range of customers will experience your product or service.

Understanding how to unite a team around a shared set of values but at the same time encourage their diversity of thought, backgrounds and experiences is where teams are at their most innovative, collaborative and successful.

Dauntless leaders do the work to help their people uncover and leverage the **strengths** of their team and find the **purpose** in their work. They work to instil and promote the **values** which drive their business and bring people together to form true **teams** who care about their individual and collective successes.

TEAMWORK—before you read on ...

REFLECT

Think about your current team:

- When has your team been at its most creative and innovative? What was it about that time that encouraged innovation?

- How similar or different to you are your teammates? (Think age, background, sexuality, demographics, cultural affiliations, neurodiversity, interests, habits, strengths.)

- What rewards and recognition exist for your team?

- What do the words **strengths, values** and **teams** mean to you?

GET CURIOUS

Ask three people in your team about teamwork:

- What are three words they would use to describe a great team player?

- What do they believe your team does to build teamwork?

- What percentage of their day do they spend on things they believe are their strengths?

- What do they believe could promote better teamwork in your team?

TAKE ACTION

Record your reflections and the responses of others.

Read the chapters on Teamwork and use these notes to help you build a plan for your team to thrive when you complete this section.

STRENGTHS & PURPOSE
[THE STORY]

Dauntless leaders help people leverage their strengths and find purpose in their work.

I lick my lips and swipe the stubby metal fork across my thigh to clean the last traces of food from it. My cam pants are black and grungy and I sweat in the piercing Darwin sun as I stand over the gas stove to cook the communal lunch. I drop the fork into the depths of my map pocket.

I splay my fingers and rub my hands down the front of my pants in a single, long swipe to wick away the sweat of my clammy palms. Now the layer of filth which has built up on my thighs is a damp layer of filth.

"Anyone want any more of this?" I call as I scoop the remainder of lunch into a mug, determined to pass it off to someone.

I head towards the vehicle where two of the team are on shift, hand over the food and hoist myself into the spare seat. I fiddle with a few dials.

"We still haven't found them today?" I ask.

"Nope."

I adjust equipment showing the junior operator through a tricky search sequence. It's Vinnie's second exercise and he pulls out his notebook to write the process down; embellishing it with annotations and diagrams so he can do it himself next time. We find the squelch of communications under encryption and note down the frequencies but fail to find anyone operating in plain, unencrypted speech.

As detachment commander I'm in charge of our six-person team. I'm yet to do my promotion courses but the Army is quick to thrust leadership responsibilities onto young soldiers.

"I'm taking Vinnie to watch the battle run," I say to Mags.

She nods and waves her hand at me to go. Vinnie and I amble away from the vehicle together with weapons slung and I wave the neat orange rectangle of a PK chewing gum packet in his direction.

"No thanks," he says as we pause in the meagre shade of a spindly gumtree. We overlook a broad valley which winds like a river through the outback station with its sparse hummocks of spiky saltbush. Grey-green leaves shimmer above us in a wispy breeze as we wait for the armoured vehicles.

"What's the most interesting thing you've learned out here so far?" I ask as I lament how unsatisfying chewing gum is compared to the smokes I've given up.

"Umm, well … Hendo's taught me a lot," Vinnie says as he scratches at his grimy neck. His nails scrape flesh-coloured tracks into the build-up of sweat, dirt and camouflage paint. "How to set up the antennae the right way, where to put them, I didn't know much about antennae before and she gets it right every time."

We chat about antenna orientation for a while and Vinnie pulls out his notebook. I watch him scribble down details and he asks me a slew of clarifying questions before he finishes his notetaking. The rumble of Armoured Personnel Carriers (APCs) echoes up the valley as they move into position. Rusty ochre plumes mark their progress against the brilliant blue sky.

We're not involved in the battle run but have chosen a spot to watch it that is out of their way. I rummage in my webbing and pull out a crumpled packet this time. Throwing the baby wipes to Vinnie after taking one for myself we rub at the skin on our faces, wiping away the layers of paint and grime and I ask, "What else?"

He flips back through a few pages and chews on the end of his pen. "Mags showed me how to find where the enemy is even when we can't force them into plain speech. That was pretty cool."

The thunder of the APCs builds to a crescendo of shifting gears and revving engines and suddenly they leap into view. The scant salt bushes are crushed under their tracks and the red earth trails in a cloud behind them. As they pass us the vibrations are intense and the sound loud and violent.

When the reverberations stop Vinnie speaks again. "I only graduated a year behind you, but it feels like you guys are just so far ahead. I feel like a total passenger."

"Why do you think I picked you to be part of the team?" I ask.

"I thought you had to take me," he mumbles, picking at the dirt under his fingernails with the tattered filthy baby wipe.

"Yeah, sometimes it's like that, but not this time."

"I'm, ah, I'm … I'm not sure."

"Well do you know how I got here?"

He shakes his head.

"Let me tell you how I joined the Army. Might help you figure out what you bring to the team."

≈

I was desperate to escape the small country town I grew up in and had decided the Navy was my ticket out. I begged my parents to drive me to the nearest recruiting office in Albury/

Wodonga an hour away because I wanted to be a Marine Technician.

When I walked in the door a stocky, broad guy in green camouflage was shuffling paperwork into a pile.

"The Navy guy is on lunch," he said after I explained why I was there. "But, don't worry about it mate, I can talk to you, it's all pretty much the same."

I stop to look at Vinnie and he hacks out a laugh. It is **not** pretty much the same.

Within an hour I'd gone from my plan to be a Marine Technician in the Navy, to signing up for aptitude testing with the Army and based on those aptitude tests, I had a range of job choices.

"You're good at problem-solving and finding patterns," said the broad stocky recruiting sergeant. "Strong communication, good English, maths average."

"Maths average!" I said with indignation.

He raised an eyebrow at me.

"Yep," he said as he spun around the piece of paper which ranked me against averages and a range of parameters.

A long list of jobs with complex-sounding names ran in a long vertical list on another sheet and he indicated with a pen from the top to the bottom.

"You can pretty much choose from anything but Psych clerk."

I looked at the list.

I wanted to seem like I knew what I was doing, but it was too overwhelming, the words too foreign.

"What do you think I should do?"

"Well I can't choose for you but let's look at the ones which suit those things we said you scored well at."

He stabbed a finger at the list. "How about this one—Electronic Warfare operator."

"What's that?"

"Well, you need a top-secret clearance, they do a truck-load of secret squirrel type stuff and it's in the top pay grade."

I shrug at Vinnie and pop another chewy in to refresh the flavour. "That guy knew how to sell to seventeen-year-olds, hey? Secrets and money?! Bloody hell, I was in."

We share a grin before I continue.

The sergeant did give me more info. I was already sold but before I signed up he told me what he knew about the job.

"One of my mates is an EW operator," he said. "She says it's all about figuring out how things relate to each other—trying to work out how people are connected to each other or what information they are sharing—uncovering the common links."

I nod at him, not sure if I really understand what he is saying.

"She says there is such a huge amount of signals and communications information out there it's like a giant, tangled ball of string where all the threads are different people, information, and connections. The way she explains EW is you're teasing away the threads and carefully pulling them out, looking at what else they are connected to and figuring out the best way to untangle them. The more strings you can follow, unwind and extract, the more you understand how they are connected. You get to the heart of the problem and can start to find answers."

〰〰

"Pretty spot on for the job, right?" I say to Vinnie and he nods.

"He was pretty spot on for me too. I wouldn't have phrased it this way at the time, but I know now I do my best work when I'm doing the stuff the recruiting sergeant described."

"I am good at following trails, exploring my hunches and sifting through masses of information. I can distil it and see all the patterns then I can find the best way forward to solve things."

I see his eyes light up as Vinnie connects the dots. "Oh yeah, Mags came to ask you how you figured out two callsigns were connected last shift."

I nod. "We've all got different strengths, mate. Everyone here adds to the team. There is no point in us all being good at the same things—we're at our best when we can harness all the different aspects of our strengths."

I watch Vinnie think for a few moments. I reach over and tap at his notebook.

"What—I take good notes?" he says.

"No, you fucking numpty. You just asked me six questions before to make sure you completely understand antenna theory. You wrote detailed notes on the search sequence. You're curious. You're willing to learn. You keep asking until you get to the answers. You're awesome at details."

"Yeah, I'm good at the details, but isn't everyone?"

"Nope. I hate them."

"You hate details?"

"Yep. Hate them. Can't stand them. Takes too much time, pisses me off, I just want the general gist of a thing and then to get on with it."

Vinnie laughs at me. "You can't actually work like that!" he exclaims.

"I know, that's why I always make sure I have someone good at details on my team. Otherwise I'll go running off once I have the gist. I need a details person to make sure we do things right, to wait for all the info and to keep me in check."

I can see him working through the times he has already done this in our team, and he smiles at his own contributions.

We turn to head back to the detachment. "I raised one other thing at recruiting before I signed up," I say. "Told that sergeant I was a bit worried about my running coz I didn't do any ..."

I smack Vinnie in the arm to get his attention. "You know what that bastard said? You'll be fine, mate—there's not much running."

Vinnie cracks up. "So, the lesson is everyone working at recruiting is a liar?" he answers with a grin.

"Well, yes." I laugh. "But also the lesson is to spend time doing work which lets you play to your strengths. I love trying to figure out the solution to something which looks unsolvable and because I love doing that stuff this job is a great fit—lets me leverage my strength."

As we arrive back at the vehicle I remind him. "It's the same for you. Being curious and being good at the details will see you succeed at this job too. We'll approach things in different ways and go about finding different solutions because of our different strengths, but we can both find a way to succeed because we both have strengths which suit this type of work. It's about figuring out how we let you do the stuff you are good at and for me to do the stuff I'm good at. Same for everyone in the team."

STRENGTHS & PURPOSE
[THE LESSON]

I was lucky to find a job so suited to my strengths. It is a privilege to not just slog it out at the job that you have but rather to spend your time doing things you enjoy and are good at.

I was able to find this match because the ADF takes a recruiting approach that focuses on aptitude and ability, not previous job experience. In a series of tests, applicants uncover their aptitudes and abilities. Potential recruits work through mechanical reasoning tests to identify which cogs will turn what way when a pully is triggered. They work through abstract reasoning tests, spatial reasoning tests, pattern recognition, analytical, situational and logical judgement tests.

The ADF attempts to predict where people will find a successful job fit based on the things they find easy, the things they show aptitude to learn and the things they are good at.

It's a very different approach compared to corporate recruiting, which mainly focuses on what you've done before—your previous work experience—rather than your ability to learn.

Aptitude testing allows the Defence Force to explore the things people will likely succeed at by measuring:

- What are the things you do well—quickly, easily, and accurately?

- How do you understand things around you—machines, cogs, levers, the world, yourself?

- What are your natural strengths?

- What do you have the aptitude to learn?

> Aptitude testing shines a light on the things you have a good chance of doing well at.

I've found a focus on aptitude over previous experience can help organisations unlock capabilities in their existing workforce and to recruit far more broadly into their teams.

It doesn't need to be formal aptitude testing like the ADF uses, but getting hiring managers to explore more deeply the things people are good at, like to do and have the aptitude to learn, allows them to recruit differently. Dauntless leaders know the value of cross-pollinating their workforces with people from outside their industry and bringing in people who think differently to their existing teams.

It is the analytics team leader who looks to recruit a communications specialist because there's no point having great insights if you can't share them in a compelling way.

It is the school which encourages students to choose subjects they enjoy, that they do well in and find interesting. The school knows those factors will see students succeed in their studies and perform better than if they encourage students to slog it out in subjects they 'think' they should do.

It is the bank who hires a veteran because they see the value of the critical thinking and complex problem solving they can bring to their team. They don't focus on their lack of banking experience.

It is the accounting firm who hires a history graduate because they value their ability to criticise and synthesise broad swathes of information into key recommendations.

It is the community services team who chooses a computer science engineer to join them. The hiring manager didn't just bin the resume as a mismatch. He scratched beneath the surface of the application to uncover decades of mentoring, networking and connecting behaviours.

> Dauntless leaders look beyond the obvious to leverage the strengths of the individuals in their team.

Rethinking the way work is done in an organisation also allows people to play to their strengths more often. Designing roles with the scope to let people pursue their strengths and interests produces some incredible outcomes.

Google famously allows their people to spend twenty per cent of their time on side projects of their choosing. Googlers can explore an idea which interests them or join someone else's project, and the strategy has produced behemoths like Google Maps, Gmail and AdSense.

This type of approach, allowing people to pursue work suited to their strengths, purpose and passions, sees leaps in innovation and problem solving which benefit the organisation and massively increases engagement and happiness for the individual.

STRENGTHS & PURPOSE
[THE APPLICATION]

When I was appointed as the program manager for a new Inclusion and Diversity initiative, I found myself with another fantastic alignment of my strengths. I love working on solving complex problems that have a sense of purpose. Untangling something knotty and finding a way through to a solution is work I really enjoy. It's even more enjoyable for me when there is a compelling benefit to solving the problem; something that makes a real difference to the world.

The organisation I was working for wanted to find a way to attract people with neurodiverse thinking to their teams and create pathways to a career for people with autism. My role was to research and recommend an approach, then design and implement a program to meet those goals.

The program had strong support from individuals and the organisation but the complexity and nuances of how to make it work successfully for both the participants and the business meant it had taken longer to get off the ground than leaders had hoped. There was genuine concern about negative impacts on participants if we didn't get the program approach just right and they needed to create something that was sustainable—not just a 'one-off'. These aims had stalled the project with no clear way forward. I was appointed to find a way past the barriers and design the path for implementation.

I spoke to every person I could find who'd run similar programs at organisations across the globe. I spoke to every leader in our business who'd expressed support for the program. I also talked to teams where innovation was important and teams where roles had strong analytical, creative thinking and detail orientations. I engaged with specialist autism recruiting firms to determine who could help us design a bespoke program to suit our needs. I made a plan, signed up our partnership with specialist agency Specialisterne Australia, agreed on generous intern salaries and we got started on delivering the Tailored Talent program.

Rather than traditional interviews and recruiting approaches, Tailored Talent candidates completed a three week, on-campus assessment program which allowed them to showcase their skills and abilities. They were given projects to deliver, robotics systems to program, bug and debug, and were introduced to potential leaders, teammates and jobs from across the whole business. Technology, Human Resources, Risk, Communications, Cyber Security, Finance, Strategy, Investment Platforms, Fraud and a heap of others wanted in. Diverse teams, leaders and people wanted to be involved with autism hiring and we were able to show the candidates all the types of work they could be doing if they joined us.

Some leaders wanted to be involved in the program because they had a personal connection and lived experience with autism. Others wanted to be involved because they needed to accelerate innovation in their teams and saw access to neurodiverse talent as a way to do that. Some leaders saw the program as an opportunity to diversify the talent in their teams by recruiting differently, while others simply wanted to give someone who'd faced barriers to work a chance to start a new career.

Whatever their reason for being involved, Dauntless leaders (and I use the term 'leaders' not as people with powerful roles but to describe the people who wanted to lead change in their workplace through influence too) were opting in.

This huge supporter base allowed me to design a program where we harnessed this engagement. Some people were keen to be involved but didn't have roles in their teams available, so they became mentors and buddies for program participants (both interns and leaders). Others recruited Tailored Talent interns into their team. These leaders completed additional training and examined their own leadership style, adapting their approach to be more inclusive and, in doing so, all their team members benefited. Other volunteers supported interns by providing future career pathways across the business and opened up networking and connection opportunities.

I knew this type of work was me playing to my strengths. I was solving a complex problem for the business—pulling together different ideas and information—and then forging a way through to a solution. But when I started to see the innovative ways that our interns, teams and leaders were leveraging strengths, I realised just how much further playing to strengths could be taken.

One intern had shown enormous strengths in process improvement during the assessment period and when he started in the risk team, his team leader spent the first week exploring this strength through conversations, task allocation and team induction. His Dauntless leader asked the intern to look for efficiencies in their team. The Tailored Talent intern observed team operations and within days he'd identified a risk review cycle that pushed data through two independent, unlinked systems that required manual data entry to move information and analytics between the two systems.

"I can write a code that does that for you," he said.

Within a week he'd written the code and when the risk team rolled out the new process, they discovered that they had saved five hours of manual, unproductive work ***every week.*** More than five weeks' worth of work every year.

He loved that he was able to create value so quickly and his team loved it too. Rather than point him to a traditional risk role in the team his leader asked him to continue to find process improvements across the whole risk business.

They were gaining far more value from him playing to his analytical process improvement strengths than they would from having him complete general risk tasks.

> Dauntless leaders look for opportunities for their team to use their strengths more often and let people chase their hunches.

Another intern had a passion for cyber security and when a vendor flagged a new update rollout, the intern spent the next three days exploring the coding and identifying potential outages and issues with the upgrade. He was able to determine with better accuracy than the vendor which fixes would work, and which would be problematic. Because he cared about the work and loved exploring the detail, this intern provided unique insights for the organisation and the vendor that nobody had been able to do before. He was promoted twice before the end of his first twelve months in the business.

Within their first months in the business, Tailored Talent interns were:

- Reviewing papers for the board.

- Designing new online programs to support staff onboarding.

- Identifying undiagnosed faults with platforms.

- Investigating unique solutions to problems which had persisted for years.

- Contributing to their teams in ways that helped the whole organisation take a more innovative approach to all sorts of work.

Each of their leaders had embraced a strengths-based approach. They had selected their new team members based on their abilities, aptitudes and preferences and then designed roles to fit the individual. Teams, leaders and interns produced truly incredible outcomes together.

> Dauntless leaders are brave enough to choose someone based on what they *could* achieve, not just what they have achieved in the past.

I realised that it wasn't just doing something people had a natural aptitude for which helped them derive so much pleasure from work, it was also the purpose behind that work which drove people to want to strive to improve their performance and be the very best they could.

I always knew the purpose behind my role in the Army but when I deployed to Afghanistan, the reality of the impact I could make really hit home. When I made the right call, I could get us out of harm's way. When I didn't there were dire consequences. What I did—and how good I was at my job—really mattered. Working hard at my trade to be the very best I could be would have a real and measurable impact on our safety.

And this was true for all my teammates, too. Each person had a role to play that was essential, important and with dire consequences if they got it wrong. I don't need a reward or recognition program to motivate me to do well in this environment—I give my all because other lives depend on it. Although work may not always be a matter of life and death, people thrive when they believe what they do serves a purpose.

You will have seen this in your work and life too.

The teacher who believes education can change the trajectory of a life. Even at the end of the term when the kids are wild and the parents even more so, he is focused. He knows his students are building their futures with every lesson.

The social-entrepreneurs and volunteers at Orange Sky Laundry (a mobile laundry service for those who are homeless or sleeping rough) who know their work is as much about the conversation while they wait for the washing as it is about getting clothes clean.

The teller at the bank who's spent decades supporting the lives of her customers. She knows that her job has so much more meaning than simply handing out money. She keeps a list of the 'oldies' who are living in aged care and who are alone. She helps them with their banking and listens to their stories. When they return home from holidays, they make a trip to the bank—not to do banking, but to show her their pictures. She sends them a birthday card every year and they come in to thank her because it's the only one they got.

The receptionist at the mental health unit who knows she may be the only person who greets the clients accessing services with a smile this month. She matches customer greetings of buoyant exuberance on arrival with an excitement of her own. She is glad to see them, glad to be supporting them, glad to have them in her life.

The palliative care nurse who finds purpose in supporting both the dying and the living during moments of trauma.

〰〰

When I was young we used to hunt grasshoppers down by the river.

In a haze of swirling pollen 'hoppers' would bounce around my dad, my brother, my sister and I. They would perch on tall thin stalks of grass in the buttery light. Long, lazy summer afternoons where the time between school finishing and dark falling stretched into a grainy blonde sepia with the smell of paspalum thick in the air.

When I was small it seemed impossible to catch 'hoppers mid-jump without squeezing the life out of them, but by my teenage years I was an expert. There were no escapees when I added new inmates to the plastic ice-cream container prison. It was a holding cell rather than permanent incarceration; though it was a death sentence for those that got scooped out and put on a hook before the afternoon was over.

I would pop glistening plump berries in my mouth and shiraz-tinted juice would run down my shirt. The blackberries grew wild on the other side of the river and when I was big enough to wade across there myself, I would guzzle them handfuls at a time.

We used to stride down to the flat, grey rocks on the bank and flick our rods into the deep still holes beyond the babbling shallows. We would wait. No chat. Just shared silence stretching out into the valley. When they were biting, we would wind in the silvery trout to flip and flop on the smooth stones. Dad would pin them with his broad, freckled hand before removing the hook to send them back into the icy alpine river.

A decade later and with an urgent call to return home, there'll be no more fishing.

I walk to our tiny country hospital hand in hand with Damo. The orderly at the station gives us a nod as we head down the corridor. In the small town I grew up in the palliative care ward is just a single room with one bed.

As I push open the door Dad teases me about my expanding waistline, insisting I can't blame it all on the baby. He's still sharp enough to spot the white flakes from a coconut slice littered across my t-shirt. I offer to smuggle some in for him, but he gestures to the tubes hanging out of him.

"Not eating too much these days," he says.

A nurse glides in and hugs each of my family who waits in that room.

"It's good to see you, honey," she whispers and pats my belly when she gets to me.

She checks the machines then leaves us. No longer rationed or on demand the morphine pumps with a muted click and whispered hiss every few minutes. Through the window I can see the tall, skinny stalks of grass swaying by the river. In the room, I watch dust motes dance in the ribbons of light from the setting sun.

I wait. We all wait.

It's familiar; waiting with our dad.

When Dad's chest stops rising and falling the palliative care nurse returns to the room bringing in a monitor. Through our tears she straps the device across my stomach and turns up the volume.

The sound of my son's beating heart fills the room in the space where my dad's has just stopped.

When you love what you do, get to do what you are good at and understand the purpose behind your work, you have the opportunity to truly connect with people in these moments.

The palliative care nurse didn't have to bring in the doppler fetal monitor. She had supported dad through his end of life care and technically that was her job done. But she knew she could make a real difference when she supported both the living and the dying. She chose to bring herself closer to us, to step into our pain more deeply and to provide us with some hope at a time that we felt hope was lost. It was so far beyond her 'job', but she did it because it was her passion and her purpose.

At twenty-two-years-old this was a formative moment in my young adulthood. Dad's nurse may never realise the long reaching impact her actions had on our family by choosing to go well beyond her duties. Actions to help our family grieve and to include my son in the story of the grandfather he would never meet.

When you are working with your strengths, passion and purpose, you can find these moments in all you do. It is where you do your best work. Where you know you are achieving and delivering at your maximum output. Where you know you are making a difference.

Dauntless leaders help their people find this purpose in their work.

ACTIONS—STRENGTHS & PURPOSE

REFLECT

What are your strengths? How do you know?

- What are the strengths of your team—how are they different to yours?

- What is the purpose behind the work you do? Do you believe this?

GET CURIOUS

Ask the next three people you interact with about their strengths:

- What are the top strengths you bring to our team?

- What strengths do you have you don't get to use often?

- What strengths do you think your teammates bring to our team?

- What's the purpose behind the work you do every day?

Consider the following:

- How well do your people know their strengths?

- Did the answers match what you would have said were strengths for those people?

TAKE ACTION

Start a conversation with your team about strengths. Have the team share the type of work they love and where they are at their best. Look for opportunities to share or split tasks differently so that people can do more of the work they love and less of the work they don't.

VALUES
[THE STORY]

Dauntless leaders encourage behaviour that aligns with the team's values and create opportunities for their team to put these values into practice.

I have just completed my final year of high school and am still a few weeks away from my eighteenth birthday when I arrive at Kapooka at the 1st Recruit Training Battalion.

Every Army recruit who is not an officer spends their first months at Kapooka learning the basics of soldiering and I'm equal parts nervous and excited as we pull in the gate. The buses come to a stop and along with about thirty other new recruits I'm herded out onto the road in a barrage of shouting. The people shouting stalk back and forth with their shoulders thrust back and perfect posture. They yell at us to hurry up, to get into formations and to pick up our crap.

Surnames are barked out in rapid fire assigning us into groups of four. I try to remember the names that were yelled out at the same time as mine as these will be my roommates for the next three months. There are no introductions, no team building, no ice-breaker activities. Instructors shout our names—we are now a team.

We run up the stairs with the metal steps clanging and ringing under the pressure of our hurried feet. We are ordered to stand outside our rooms. I look straight ahead, shoulders pressed against the wall in the long, pale green hallway.

Four men and two women stalk up and down the hallway and take turns shouting at us. They swagger and stomp; getting right up in the face of recruits with spittle flying as they dress them down for their haircut. Or clothes. Or posture. Or expression. Or looks. It doesn't matter. They will find something. As one bellows the final words of a tirade, the next starts.

I glance out the side of my eye at one of the women I'll be sharing a room with to see if she is as petrified as me. The shift of my head draws the attention of a passing Corporal. He brings his face just inches from mine as he starts yelling at the volume, intensity and fury of Nirvana's 'Smells Like Teen Spirit'.

"WHAT THE FUCK ARE YOU LOOKING AT?" he roars.

"I, I … nu-nothing, sir."

He takes a breath and I shudder.

"I'M NOT AN OFFICER," he thunders, gesticulating wildly and slashing the air with his hand like it's a knife.

"DON'T CALL ME SIR. I WORK FOR A LIVING."

I feel like I'm in a cartoon where he's shouting so violently it blasts back the hair back from my face and I try to weather the storm. The image is so ludicrous I worry that I'll start giggling but I'm also a bit terrified.

My mind races and tumbles for a response.

What does that mean anyway 'I work for a living'?

Am I supposed to respond?

If I don't call him sir what am I supposed to call him instead?

Oh my God, I don't want to get into trouble.

I frantically look at his shirt for clues. I had reviewed the Army rank system before we got on the bus.

OK two stripes … what the hell does two stripes mean?

What rank is that? I thought you had to call everyone sir!

He's still standing there right in my face.

His eyebrows draw together in a scowl and I can see his nostrils flaring as he takes a breath ready to bellow at me again.

Say something, Rach. Say something. I think.

"Yes, sir," I blurt out again.

"ARE YOU FUCKING DEAF, RANTON?" he explodes.

A few of the other corporals come over to join him and they share shouting duties for a minute, all karate chopping the air with vigour to drive home their points.

Lucky for me one of the other recruits starts sniggering at my misfortune down the hall. The head of every Corporal swivels and like a pack of aggressive velociraptors from *Jurassic Park* the section commanders turn and swarm to attack the new victim.

VALUES
[THE LESSON]

This is the picture most people have of the military and it is a true one (or it was for me in 1997).

But it is true only for the first few weeks.

When you arrive at Kapooka the Army works hard to jolt you out of your civilian world; the feedback is harsh and the culture one of unrealistic expected perfection.

It's a shock tactic to help you shake off your old world and start living in a new one. But the behaviour of those leading soon changes.

Even while you are still at Kapooka at Basic Recruit Training, within weeks this command and control approach to leadership shifts. Recruits are quickly empowered to take more ownership and make more choices. Values come to the fore and soon are both the principal and the simmering undercurrent of every lesson.

Weapons training.

Drill practice.

Navigation.

While you are being taught technical skills, you are also being driven towards the behavioural expectations around the Army values of Courage, Initiative, Respect and Teamwork.

The Army does not just shout the single words at you and expect you to live these values. Instead, Kapooka helps to turn young Australians into soldiers who uphold the values of the Australian Army—in every circumstance—by providing situations, feedback and learning where participants can demonstrate and integrate these values into how they work.

Every scenario provides an opportunity for new soldiers to deliberately practice these values until they become etched into their DNA—they become a part of who you are and how you do things—and everything you do reinforces the concept that your values are the Army's values, and the Army's values are your values.

They did such a good job embedding these values into every soldier I still see them in all my military mates, no matter how long they've been out. I can rely on my veteran mates to be courageous, to get shit done without asking, and to be on my team. Courage, Initiative, Teamwork.

My veteran mates also show respect to others. Even if that respect comes with lots of shit-talking and swearing, there is a genuine underlying respect for people in all that they do. Soldiers see humanity at its best and worst. We are there to support the community during floods and fires. After disasters, in conflicts, as peacekeepers, we work side by side with local people. During conflict, they will rush to help save the lives of those who were shooting at them only minutes before.

Soldiers know what people are capable of and see the raw emotions of loss, failure, grief, elation, freedom, and hope that exist in so many of the situations they find themselves in. It makes us respect what people are capable of. The good that humans can do and, at the same time, the terror and destruction we can cause each other.

I believe the military does such a good job of aligning the values of its people and the organisation because they create opportunities for soldiers to practice the values.

Take the Australian Army value of courage. New recruits are asked to do things that require them to show courage daily. Running, jumping, and shooting weapons. Bayonet assault courses, navigation, and marching with heavy packs. High-wire obstacle courses, drill, and endless burpees. Having to throw a grenade just after the bloke before you was dragged back into the shelter because he didn't throw it far enough. These are all opportunities for you to practice being courageous.

The self-doubt that simmers in many recruits as you worry that you won't be able to keep up the relentless pace of it all is also a chance to show your courage.

Every test of your courage occurs in public with all your peers looking on. Scenarios are designed to make sure recruits learn how to find their courage and test their fears.

Terrified of heights? There is nowhere to hide it as you slowly shake and hesitate your way through the high-wire obstacle course and convince yourself to jump, ziplining from giant towers—all in front of your whole platoon. You're allowed to be scared but you're expected to push through the fear and attempt what is asked of you.

Those mean-faced shouty section commanders from your first few days will instead beam like proud parents as they praise you for your efforts, your improvements, and for doing what you think is impossible.

Terrible at running? You'll be asked to do it anyway. Your awful times yelled out for all to hear and the entire platoon acutely aware of exactly what position each recruit finished in. You are expected to keep fronting up and keep giving your all.

Your fast-finishing peers will come back along the track and run beside you to encourage you to push hard for the final section while they cheer you through the finish line.

Rubbish at maths? You're still expected to show up even when you know you will struggle to find the answer during a navigation lesson when asked because you flunked maths almost every year at school. Your instructors know this but you will be called upon to provide answers in front of the group anyway.

The soldier beside you will show you their answers and mutter explanations quietly as you work through the problems until you can do it for yourself.

At every turn you are pushed to deliberately practice doing what you find difficult and to show your courage.

And at every turn, you will find your leaders and teammates shouting encouragement.

As Nelson Mandella so beautifully said: Courage is not the absence of fear, but the triumph over it. The brave man is not he who does not feel afraid, but he who conquers that fear.

I like to think that he meant women, too.

Again and again in the military you are asked to face your fears. You ***practice*** being courageous.

All of the Australian Army values are rehearsed and practiced until they become a habit, and this deliberate practice plays a critical role in bringing the Army values to life.

Leadership guru Simon Sinek recently released a video berating organisations for sprouting their values but giving their people no real way to live them.

What he called out is that organisations love to make their values single words—things like integrity, respect, innovation—but organisations don't help their people figure out how to put these words into action. People know the vibe of these words but in a work context they often can't put their finger on exactly what is being asked of them.

Because you can't **be** integrity, or **do** respect.

Sinek suggests an approach to take these words down off the wall and put them into action. He states that "values need to be verbs".

'Always tell the truth' is a value people can act on—'honesty' is much more vague.

The Australian Army has been doing this for decades by creating an environment where its recruits turn values into verbs through practice. Outside of the military there are ways to turn values into verbs too.

VALUES
[THE APPLICATION]

During my time leading retail bank teams, I ended up finding my niche in turning around the performance of branches and regions which were struggling with customer service and sales.

When I started work with the team in Toowoomba the branch was ranked in the bottom five for performance in Queensland. Over the next two years as we worked together, I found my feet and figured out how to translate my military skills into the civilian world. We started to climb the ladder.

Before the three-year mark we'd built a culture we could all thrive in. One where we were focused on our customers and our outcomes and one where we helped each other to succeed. We ended that year the number one branch in Queensland for customer and financial outcomes.

I loved the thrill of turnaround. Each new branch team I joined, and each new region I took over, turnaround was the task given to me and I was happy to receive it. After the first one, I knew what to do and we could do it so much faster.

What I found when I arrived at each of these 'underperforming' teams that needed to be turned around was they were full of incredible, dedicated, talented people who just weren't clear on what they were trying to achieve (no clear mission) and how they made a difference (knowing their values, strengths and purpose).

≋

Our team gathered in the break room before opening the branch.

"We're going to spend our meetings this week talking about our values," I said.

One of our top performers took a breath ready to start sprouting the values at me and I placed a hand on her shoulder.

"Before you tell me the values—because I know you all know them—what I mean is, I want us to talk about examples of each of our values. I want us to agree what type of behaviours our team wants to hold up as the best example of us living those values."

A few looked at me with a blank face, my top performer sat forward ready to be the best student in the class.

"Okay—so our first value is integrity. Let's hear all the ways, the things we do, that show we are living the value of integrity here at our branch. All ideas are good ideas. They'll all go up on the board, and then we'll decide later which best describes us."

"So, what does integrity mean to us in our team?" I asked them.

People called out their ideas, I wrote them all down.

"Not stealing the money."

"Doing the right thing by customers."

"Coming back from lunch break on time."

"Telling the truth."

They all went on the board.

"It means replacing the toilet paper when it runs out and not just leaving it for someone else."

I smiled. It was interesting to hear the things that got under people's skin that they hadn't found a way to express before.

"It's paying attention to the cash when you're on the teller line."

"No bitching."

Some ideas weren't exactly integrity, and that was okay. Some were positive examples of living the value of integrity, a few were what we thought integrity was not. I asked the team to switch their focus to include more of these examples.

"What are examples of *not* acting with integrity?"

"Hiding mistakes."

"Calling in sick on a Friday when you're not sick."

"Hoarding the best stationery."

"Avoiding difficult customers in the queue."

"Claiming credit for someone else's work."

The 'bad' examples came thick and fast. People don't often get a chance to point out the perceived slights of their co-workers or how they felt about inequities in the workplace in an open environment.

We spent lots of time talking about this range of behaviours we didn't want in our team. We talked about how they made us feel when others acted in this way and how it hurt our team when we did these things.

It was interesting to see the themes that came through on both sides of the equation and we chose three of each. Three examples we agreed

were acting with integrity and three examples
we agreed **were not** acting with integrity.

For those who like the details, we landed on:

- We do what we say we will.

- We always think about the customer's needs first.

- We are honest and tell the truth, even when it is hard to hear.

And

- We don't hide our mistakes.

- We don't leave the shit jobs for our teammates.

- We don't undermine each other.

This exercise gave us a common language and clear descriptions of how we would demonstrate integrity in our team.

It took the single word and turned it into not just one verb but a whole raft of actions that we agreed were the embodiment of that value.

I rarely had to call out bad behaviour or discipline people about how they were doing the wrong thing again. The team did it with each other.

ACTIONS—VALUES

REFLECT

Review your team's values and explore:

- What behaviours are the norm in your team?

- What would an outside observer notice about the feel of your team if they spent the day in your offices?

- How does your team work together and do they get any opportunities to deliberately practice your values?

Aim for sentences over single words—'When faced with a choice between the easy way and the right way, we choose the right way, no matter how much more difficult or expensive it is' rather than 'Integrity'.

GET CURIOUS

Ask three people in your team:

- Which is the most important of our values? Why?

- What's an example of something a teammate has done that shows they are aligned with our values?

TAKE ACTION

Work through the activity on values with your team.

Discuss examples of what demonstrating this value is and isn't as a group. Decide what the best examples are for each value and capture these/display them somewhere prominent.

TEAMS
[THE STORY]

Dauntless leaders acknowledge when people do the right thing and behave in the right way AND when people do the wrong thing and behave in the wrong way. They are right there in the thick of it with their team—sharing experiences, rolling up their sleeves and providing support when things get hard.

On Exercise in the Northern Territory

"Keep up the speed as you head out," says the dumpy ordnance bloke as we drive back out into the main exercise area in Tennant Creek. He taps on the driver's side door to get Sammy's attention.

"Seriously, you've got to go fast. Gun it!"

"Cool, mate," says Sammy as she gives him a lazy smile. "No probs."

We are all feeling relaxed and happy after our first shower in two weeks. Although it was just cold water dripping from a canvas bag behind a strip of hessian, it felt like a luxury day spa. The scent of soap on all our bodies is a welcome reprieve after the funky unwashed mess we'd been when we'd arrived into the admin area.

From my seat in the back of the Land Rover I peer around the edge of the vehicle and see the red bulldust is deep and wide where tracked vehicles have chewed up the dirt road.

Sammy keeps the gears low and revs high as we race down the track and around the corner, but the dust is a fine as talcum powder and as deep as our waists. It acts like an oil slick and the vehicle pivots, my view in the back now showing me the paddocks surrounding us as our rear end sways out.

Sammy tries to correct, we fishtail and the diff thumps on the road as we are dragged into a deep rut. The vehicle slows down as Sammy lets off the accelerator and we rain down profanities from the back seats as the Rover grinds to a halt.

We all get out of the car, give Sammy hell for a few minutes then grab our recovery gear. I loosen the wingnuts that keep the shovel on the bonnet and start to dig.

The dirt is so fine it plumes into the air and floats around us as we try to move it. Each shovel-full causes a cascade of loose grains to fill the hole I've just dug and we make as much progress as people bailing out a sinking cruise liner with a thimble. There's no other option though, so we dig.

In the beating sun we shuck off our outer shirts and work in our singlets. I wipe my forehead on my sleeve and my sweaty palms on my thighs. We're blocking the road. Every vehicle passing us is revving and speeding, trying to make sure they don't get stuck too, and I get a violent face-full of grit and dust. I start retching and snorting from the dirt enema that's been forced up my nose.

"That must be a new record," I say as I climb back into the vehicle. "We got bogged four minutes after our shower. Four minutes of being clean," I say, gesturing to my shirt and pants covered in grime and screwing up my face.

"You're also drenched, like ab-so-lutely drenched in snot," says Sammy helpfully.

"Wouldn't have been there at all if you knew how to drive," I retort.

Sammy shrugs and then smirks at me. "Whatever. All of you would have got bogged there. Anyway, it was worth it to see you deepthroat most of the dust in the operational area."

We laugh. Because what else can you do?

On a promotion course in the Victorian bush

I twist my body trying to find a way to carry the giant hoops of barbwire without them digging into me. I settle for a spot where it doesn't stab me too much, sling my weapon and I set off up the hill.

It's my third trip lugging wire, star pickets, and tools from the stores point at the bottom of the hill, to the top. I trudge under the weight of the bulky, awkward load and pass other section members, moving at the same steady rate, coming back down to get more gear.

I hear my teammates before I see them; dirt and gravel crunching under their boots and their weary footfalls landing heavily. It's well past midnight and the heavy cloud cover makes the already moonless night near pitch black. In the seconds where we are close enough to see each other in the dark, we raise our eyebrows as a form of acknowledgment and greeting, all too tired to bother speaking. I nearly stumble into one of my teammates who has stopped mid-way up the hill and is dumping his gear in a haphazard pile.

"This is so fucking stupid," he grunts out through gritted teeth. "We're never going to get all this shit moved before sun-up."

He lets his body thud to the ground and while he sits in the dirt looking up at me, I remember just how young he is.

"That's the point, Tommo," I say quietly. At twenty-six I am wise and grizzled.

"They know they've set us a task too big to achieve. They want to see how we'll react when faced with an impossible, seemingly pointless mission—do we give up or keep at it?

They're just trying to make us exhausted and deprive us of sleep so that we're pissed off and snapping at each other."

"Well it's fucking working," he spits.

"Come on you lazy bastard," I say as I help him back to his feet. "Don't let an old lady beat you."

In Afghanistan on patrol

We are five days into the patrol when I realise I can wait no longer. I pick up a shovel from the nearest car and saunter up to the guy on picquet pretending everything is casual.

"Just heading out for a bit," I say.

He grins at me knowingly.

No matter what 'business' you are getting up to, on a patrol people need to know where you are and what you're doing at all times. The guy on picquet knows exactly what I'm up to; you only go outside the 'harbour' with a shovel for one reason.

I walk down the creek line to give myself at least the illusion of privacy and I dig a hole. I hike my body armour up, shuffle my clothes out of the way, drop my trousers and try to clear my mind.

Within minutes a group who have been out on a foot patrol into the nearby village start walking up the hill towards the vehicles. They are headed straight for me, but after waiting five days it is too late for me to change what I'm doing, and they can't change direction.

I wave at them, blushing crimson.

"Hey guys, what's up?"

To their credit, they wave back like it's no big deal.

"Hey, Rach."

In Russell Offices, Canberra

The *Get Smart* theme song plays in my head. Every morning, when I get to this part of the building it plays in my head as I nod to the guard, swipe my pass and walk through the first barrier.

Da-da-da *pause* swoosh

The sliding doors open in time with my imaginary swoosh and walk through to face the next barrier.

I swipe my pass again, and the next door opens ... da-da-da *pause* swoosh

I step forward into the airlock between the next two pairs of sliding doors ... I lean forward to have my iris scanned ... da-da-da *pause* swoosh

The last door opens, and I picture myself as Maxwell Smart in the opening sequence of every episode walking through all those clanging, sliding doors swooshing closed behind him. I'm not wearing a suit though. And my shoe is definitely not a phone. All phones must be left outside the building.

I join my team at the cluster of desks in the analysts' section. Although it's been years since our exercise in Darwin, I've found myself back in a team with Sammy again. Our section is in the middle of the vast open-plan space which houses the Defence Signals Directorate. The far wall is covered floor-to-ceiling in huge monitors scrolling through maps, news, and signals information. Motley mixes of Army, Navy, Airforce and civilian personnel sit in rows and rows of desks configured into small teams. Everyone wears headphones.

Leno swans in from the break room with a single cup of coffee.

"Oh yeah jack ... um ... no thanks," I mock, using a term with particular, and uncomplimentary, meaning in the ADF.

"Don't worry about us, Leno, take a break during the busiest time of the shift why don't you," drawls another team member cynically.

"Piss off, you idiots," says Leno with a laugh.

"Jack much?" says Sammy, giving Leno the finger with a big grin on her face.

Leno lifts his hands up. "Alright, alright," he says. "I'll make sure I do the full catering job next time. I'd hate for any of you fat lazy pricks to accidentally get some extra exercise by getting your own brew."

In the barracks at 7th Signals Regiment

Mellie flicks his cigarette butt into the ashtray.

I watch its arc as it spins end over end towards the tarnished silver trough. There's a large sign above the ashtray with some custom drawings to help soldiers understand the purpose of the device. I snigger at the shapely texta-drawn bum that sits alongside a huge, veiny, hairy, highly detailed drawing of a cock and balls inside a circle and backslash 'no symbol'.

Butts only explains the lettering.

"Ahhhh, I worry about the day I'll find an anatomically incorrect cock and balls unfunny," says Mellie.

"Anatomically incorrect?" I say puffing out my chest. "My dick's that big."

"Had to use a pair of tweezers to find mine this morning," says Mellie squeezing the tips of his fingers together in a tiny pinching motion. He gesticulates near his crutch to ensure I can visualise it all. "Can barely piss over me own balls."

I snort into a fit of laughter.

~~~

TEAMS
[THE LESSON]

Teams don't come together by accident.

In the military, teams are constantly changing. They form for an exercise or operation then disband when they return. They restructure with new members after some are posted away and others arrive to take their place. You may work at the same unit as someone for years and years but never end up in a small detachment with them. Other people somehow always end up in your team, no matter where you are located or what role you are posted to. One thing you can rely on, your team will change.

Both team members and leaders have key roles to play in this constantly changing environment. When a strong culture of safety, honesty and trust exists, team members do the bulk of the heavy lifting when it comes to team building. In cultures that support a 'team above self' orientation like the military, team members do a lot of self-regulation.

You can see it when Leno returned with only a single coffee (around here we get everyone a coffee when we get one for ourselves).

You can see it when we had to dig ourselves out of the bulldust in Darwin (around here we all pitch in to get us out of trouble).

You can see it in Afghanistan (around here we all have to shit in a hole in the ground in public, so it's no big deal, you're one of us, you weirdo).

Dauntless leaders are quick to leverage the team culture to create team norms.

American psychological researcher, Bruce Tuckman, published in 1965 his 'Stages of group development' theory, outlining the four phases of forming, storming, norming and performing that teams progress through in order to grow, face challenges, tackle problems, find solutions and deliver results.

That cycle of forming, storming, norming, and performing is a constant in the military. With ever-changing teams you are always adjusting to new members, re-establishing boundaries and learning how to work best within the new group. Constant change means teams get good at moving through the cycle quickly when leaders use the forming stage to build team culture.

Multiple times a year your team will change, so multiple times a year people are navigating their way through challenges, growth, and problem-solving while building new relationships. Multiple times a year, leaders have the opportunity to leverage this change to strengthen their team.

As a result, military people have learnt to find common ground, break down barriers, and make new friends fast. Share a beer with a veteran and you'll see how soon they look to find your commonalities (where you lived/grew up, sports, hobbies, who you know, swearing, jokes—they will rattle through the lot seeking a connection as fast as they can). Once the connection is made, they'll want to spend time on it, digging into the details to bring you closer together through this shared experience.

It's not just this hunt for commonality that helps military teams through the stages of group development. New team members take cues from their peers and watch how the team interacts with each other to help them understand the type of team they are joining. They're looking for answers to questions that will tell them whether 'team' or 'self' is important:

- When something goes wrong, does the team pull together or point fingers?

- When things are difficult, do we help each other, or do we fight our own battles?

- What gets recognised around here—behaviours? performance?

- Do we include people, get close to them, make them one of the gang, or is it hard to break into the clique?

- Does the group self-regulate acceptable behaviour, or rely on the leader?

- Is it fun to work here?

People are asking themselves these questions when they join new civilian teams too. Work teams, sports teams, community groups.

> Think about the last time you joined a new group—what questions were you looking to answer so you could decide if this team was for you?

Dauntless leaders work hard to help teams build their answers to these questions in three key areas:

- Enabling shared experiences.

- Recognising behaviours.

- Refusing to tolerate unacceptable conduct.

Enabling shared experiences

Within the military teamwork and trust is accelerated by the context in which they operate:

- Prolonged, challenging experiences (like field exercises) where team members gain significant insight into the ability of their teammates to perform their tasks and their motivation to do so.

- Environments of risk and uncertainty (like operations) where individuals are reliant on the capability of others to do their job well.

- Structures which push decision making down the chain of command (like mission execution) where clear direction is provided by senior leaders, but execution is entrusted to those on the ground.

'Remember that time when ...' is the most common sentence starter you'll hear in any conversation between two veterans catching up. Talking about all those shared experiences reinforces the depth of the connection that you had at that time, and reminds you of how close you are or have been in the past. There is pleasure in the shared experience even if it was difficult. It brings a smile to your face, sharing a secret that only people who were in that team at that time know.

It's guaranteed that every time I catch up with Sammy we talk about the time she got us bogged in bulldust five minutes after having a shower and the slog it took to get us back out. She is always quick to remind me of my snot attack too.

Shared hardship and shared experiences binds military teams together, and it can do the same for all teams.

I know of a high-performing, highly successful National Rugby League team who deliberately creates a shared experience to help forge their team together in the preseason through a three hour rope run. What's a rope run? Get one giant, fat rope from off a ship, explain that everyone on the team must be touching the rope at all times, then run. It's not right for every team environment, but you can see instinctively that the purpose is to create the same type of tired, exhausted, 'this is pointless' feelings that we experienced on the promotion course when moving stores, allowing leaders to quickly determine how people act and react under that pressure.

In this environment, training staff observe who will dig in and keep going, who talks up and encourages others and how the team organise themselves to complete a challenging task. They see who gets snappy when under pressure and stress, and who looks for the chance to turn hardship into fun. They're not looking for who's physically the strongest but rather how people **act and react** under these conditions.

> Dauntless leaders are not afraid to observe their team in challenging circumstances rather than rushing in to solve problems for them.

Knowing how individuals in your team respond to challenges helps you determine what role they'll play best in your group, and what different types of support people might need when times get tough. Team members learn about each other too, and leadership from within the group—not from who holds a leadership role, but who is revealed to act like a leader when times are tough. And they'll talk for years about that bloody rope run.

Those experiences become part of the folklore of teams. They become the stories team members tell and believe about themselves; the stories of who they are and what they can overcome together. Those stories become the fabric of what it means to be in 'our' team.

Shared experiences don't need to be physical, but they do need to be challenging in some way to help forge a team's identity. It is at the point of challenge, crisis and purpose that teamwork is stimulated and grows.

A working group thrown together urgently to get the system back online after a power outage will talk about that shared experience for years to come. When they cross paths with the same people on new projects, this common story will be how they reinforce their bonds and show others in the group that they trust each other.

A crew who works to support their community through flood, fire or illness will galvanise around those stories, regaling new members with details to help the group to describe who they are and what they do.

A team working on the busiest day of the year to pump out a massive session together will likely finish their day equal parts exhausted and exhilarated if they have spent the time lifting each other up to make things fun and exciting during the high tempo day.

Specific team-building activities work but where you create a legacy story is when what is at stake is real. Most organisations find themselves with these types of challenges more regularly than they probably realise: customer failures, product issues, power outages, Boxing Day sales, recalls, natural disasters, mergers, IT crashes, complex restructures, security breaches, redundancies, acquisitions, booms and busts, are all opportunities that can create legacy stories if they are leveraged correctly by leaders.

Dauntless leaders use challenges to build their team.

Dauntless leaders are deliberate about who they select to join a group to face and solve challenges. Dauntless leaders are looking for opportunities to build stronger connections and deeper relationships across the team, and at the same time to enrich the story of who they are as a team.

Dauntless leaders look at challenges as opportunities for their teams grow.

Shared experiences show up in odd places in the military. I'd hazard a guess that taking a crap is not the first thing that usually comes to mind when you think about shared experiences.

The thing is those day-to-day activities that we all have to deal with as humans proved great commonality in the Army. In the field, regardless of your role, rank, gender, capability, age or experience, when you are out bush or on operations, everyone has to crap. You work side-by-side with such a lack of privacy, in such close proximity and with such intimacy in those environments, that your ablutions become a normal topic of conversation and just another task on the 'shit to do' list.

That closeness and intimacy of living and working with people 24/7 pushes way beyond things like this and deep into the details of each other's lives. You know all about your teammates' kids and relationships, their hopes, fears and ambitions. The sheer quantity of time spent together means you see people at their best and their worst.

You're away from your normal support networks and families, so you become that for each other. At the same time, there is little anyone can do to change the circumstances you find yourself in, so you find ways of supporting each other and coping through these honest, raw exchanges, usually with a sense of humour.

Jokes are made to bring people into the fold and while there is a fair amount of good natured ribbing, jokes are rarely at the expense of others and far more likely to be the observations that bring you together—like crude drawings to remind you of the purpose of an ashtray.

Civilian workplaces are not the place to share your daily ablution schedules to create a sense of team but you can work to increase the intimacy in your teams by providing them with the chance to build common experiences together and the space to share the details of their lives.

Recognising behaviours

How often do you reward people for how they act, rather than what they have achieved?

It's hard to measure teamwork and because it's subjective, it is often overlooked or avoided when it comes to recognition. You can't tick boxes or take a defined action to confirm teamwork has been completed for this month.

It's much easier to justify an award given to the person who made the most sales (as you can see that on a spreadsheet at the end of the week) than it is to hand out an award for best team player (as you have to dedicate the time it takes to observe how people act in the team and then subjectively choose who is the 'best').

Despite the difficulty, Dauntless leaders find a way to recognise those behaviours that contribute to positive team outcomes. They are not put off by the lack of data.

Dauntless leaders are not afraid to make calls
that are subjective.

I found out many years later the reason I was presented with the leadership award on that promotion course where we moved stores in the dead of night was because of moments like the one on the hill. It was my few quiet words to encourage a teammate to keep going—not my own completion of the tasks—that was recognised.

As a leader, you've got to be close enough to the action to observe and catch these moments. You've got to spend time with your people as they complete the day-to-day actions of their roles to see who does the things that help your organisation pull together as a team. When you are small, this is easy. You're part of the team and know who puts their hand up first to take on extra tasks and who checks in to see how decisions will impact on other parts of the business. As you grow, this gets harder to see. You've got to make the time to observe how people act and decide how you call this out.

The best team players do the things that help everyone succeed. Their behaviours and actions—big and small—are performed consistently over time, always pulling towards the team's success and not necessarily their own. They are the teammates who organise a lunch outing. The ones who collate the weekly report. They are the teammates who make the templates and share them. The ones who ask, 'who needs help today?' It can be easy to overlook the amount of work someone in the middle of the pack is doing so that others can soar. You have to make time to see it.

Failing to recognise behaviours sends a message to your team that only outcomes matter and can lead to people focused on their own outcomes, sometimes at the expense of the team.

Refusing to tolerate unacceptable conduct

Home-grown Aussie tech juggernaut Atlassian made international headlines when they changed their performance review measures in 2019 in an attempt to weed out the #brilliantjerks loitering in their organisation.

Atlassian wanted to find a way to stop what they called 'brilliant jerk syndrome'; where someone who is technically talented is held up as a top performer, even if the way they achieved their results is at the expense of others.

They did this by shifting their performance criteria focus away from individual outcomes and towards the values and behaviours which make great teams and great team players in their business.

In this approach, it's more important for me to 'do the right thing by other people' and to 'proactively collaborative' (as these are measures on my performance scorecard) than just smash out my own results. My results still appear, but I'm assessed against a framework that considers if my success makes others fail.

It's an interesting attempt to put measures behind teamwork, because Atlassian knows that when you get your jerk-management wrong, it'll destroy your team in all sorts of insidious ways.

I once worked in a team with a jerk who, on a scale of 'one (great bloke) to ten (toxic dickhead), was an eleven! This wasn't a case of an OK dude behaving badly because something was happening in his life, but a consistently nasty, underhanded, uncooperative jerk.

On the surface, he was always a performer. He always hit his targets, but he also spent his time bullying and intimidating his subordinates and peers. He seemed to genuinely enjoy backstabbing, spending a ridiculous amount of time bitching about others while forming and reforming 'in' and 'out' groups based on anyone who dared to raise a different opinion to his.

He would use any means necessary to win on a scorecard and used gossip like a weapon to take down anyone who might produce a result that outstripped his. I never once saw him forgo a short-term bonus for any of his publicly stated commitments to delivering on our values like teamwork, integrity or innovation.

It was about character, not about a specific situation.

It's important to differentiate between these when you're thinking about jerks in your team.

> Stop and ask yourself: Is this jerk behaviour
> or someone who is a jerk?

Good people can sometimes act like jerks if stuff is happening in their life.

I've been there, and if you're honest, you've probably been there too.

Dauntless leaders ask insightful questions in genuine conversations to determine if they're dealing with someone behaving like a jerk due to a situation or if they are behaving like a jerk due to their character.

Situational jerks need some Radical Candour.

Character jerks need urgent removal.

And Dauntless leaders are quick to act on both.

In the military, jerks get the 'you're being a dickhead' candid chat I'd received. But when their behaviour doesn't change, the group will act.

Jerk behaviour will be called out by every member of the team, every time they see it. The military even has a helpful phrase to let everyone call out jerk behaviour: **'Don't be jack.'**

There's a few origin-stories of the expression 'don't be jack'. Most include acknowledging a heritage from the British military or Navy with the meaning behind calling someone 'jack' being something close to *someone who only acts in their own best interests, even if helping others would require bugger-all effort from them.*

Finished a task and now sitting around doing nothing while your teammates are still working? That's jack.

Taking the least weight in your pack and letting a teammate carry more than their fair share? That's jack.

Did a lazy job cleaning group gear and now the whole team has to stay back and redo your work? That's jack.

Made a brew for yourself, but not for everyone else? That's jack.

Being called 'jack' is a serious insult.

The power of 'jack' is it is a common language for everyone in the Army, regardless of trade, rank or status. People understand the connotations and inferred meanings which are embodied by that single word.

In the military, acting in the best interests of the team always comes first. It's why Leno was expected to ask everyone if they wanted a coffee before he snuck off to make one himself and why he was called out on it by the group when he didn't. Team before self. Leno knew this - it's why he capitulated (quickly and with humour) to the feedback from the team.

Dauntless leaders spend time calling out behaviour that is unacceptable in their team *and* rewarding the behaviours they want to see. Doing one without the other is not enough.

TEAMS
[THE APPLICATION]

Our shared experiences become the stories of our teams. The stories we tell ourselves about who we are can shape long lasting legacies.

In a story credited with still influencing the current Silicon Valley mindset and ethos around teams, HP co-founder, Bill Howlett, came to work one weekend to find the equipment storeroom locked. He smashed open the lock with a fire axe and left a note on the door insisting it never be locked again because **HP trusts its people.**

Demonstrating trust for a team is a key value across many of the current Silicon Valley giants and they all point back to this story as one that created this sense of team trust that permeates the area.

There is also a reason we look to sports teams to help us understand teamwork at its best. All sports have these legacy stories that teams tell themselves about what it means to be on that squad. I have played sport my whole life—netball, dancing, wakeboarding, hockey, triathlon, AFL, tennis, basketball—so long as it's not golf, I'm interested in playing.

These stories are often born from key moments. It's kicking the goal after the siren to win the grand final, of the magical intercept that turned the game at a critical point of tension.

But those big stories are often created by small actions. When I think back on those teammates I loved playing with most, it's clear that they were the people I could always rely on—the people doing the 'one percenters'. The players whose actions benefited the team but most times didn't translate into personal glory. The work that would never appear on a stats sheet but made all the difference to being able to trust and rely on the person playing by your side.

The teammates doing the things that contribute to the outcome for the team, but are rarely measured. Those teammates who:

- Are vocal and encouraging, they are a constant voice on the field, keeping the team focused and positive.

- Run dummy lines, with intensity and frequency, so the defence has to provide cover for them. This person knows they won't get the ball or score the try. They won't throw the winning pass. But by running the dummy line they create the space for someone else to do those things that help the team succeed.

- Make a lead and are always an option.

- Cheer and hug, running to pat their teammate on the back for a good tackle, big hit, rejection, shot, score, defence or throw.

- When they know their opposing number is covered, they move to take up space to prevent the other team's attack.

- Focus on the synchronicity of the group. They hold back an inch on their high kick so that every squad member hits the same angle. They forgo pursuing their own perfection for the perfection of the team.

- It's the guy who buys everyone matching sweatbands for the grand final in D grade volleyball.

- It's the woman who kicks to the post knowing her teammate will come through on the right angle to score.

- It's the teammate who chases the breakaway knowing they'll never match their speed but hoping to force the opposition to score at the edges.

When we recognise and encourage these behaviours in our team, we all benefit. When we recognise those doing the work that doesn't always translate to the scoreboard they feel valued, acknowledged and continue to do these things that make a difference to how people feel as part of the team.

I once worked in a team where the leader focused on 'one-percenters' as both a weekly and major award of the year for our team.

Each week he asked us to share the stories of someone doing work which made the team stronger. He pushed and probed us on the types of stories we told, steering us away from the traditional numbers, deal sizes and sales outcomes until every week we were all sharing stories about the small things that made a big difference.

The lending officer who was meeting a couple to help them with their first home loan. As their small daughter drew in the corner her parents talked of the property they had found and hoped to make their own. The beautiful jacaranda tree in the front yard. The bold red front door. At the end of the meeting the family left, and the lender found the daughter's drawing of the house they'd described—purple jacaranda in the front year, bold red front door. He kept the drawing and on settlement day delivered it framed to the family to welcome them to their new home.

The teammate who got to work five minutes early every day to prepare a daily thrill for his teammates. When the rest of the squad walked in the door they would find their favourite coffee, tea or cold drinks already at their desk. The delight on their faces was only surpassed by the delight on his at their pleasure.

The teammate who always put her hand up to help customers when they came in to advise of a deceased estate. She knew she could make at least one task during this tough time easier. No matter how many tears she'd already shed with other families, or if it was well past the end of her shift, when someone walked in to share their loss and ask for help to navigate the tricky waters of estate management she would stand up and say 'I want to help' with a compassionate heart.

Our Dauntless leader was brave enough to focus on the things he believed mattered most, and to have faith that the scorecards would take care of themselves. Other leaders I'd worked for in that role had been relentlessly focused on outcomes and numbers. Our results had fluctuated up and down as teams grew tired of pushing and pushing to achieve the targets. Under this new leader our results steadily increased and rarely dipped. Our outcomes were ever increasing, and over time they grew to far exceed the old highs by large margins.

Spending our time focused on the small things that made the experiences of our people and our customers exceptional improved our results much more than focusing on the numbers.

ACTIONS—TEAMS

REFLECT

When was the last time your team faced a challenge? What happened and how could this have been leveraged to create a shared experience?

What stage of Tuckman's 'Forming, Storming, Norming and Performing' cycle are your team at now? How good are you at moving through the cycle when team members change?

GET CURIOUS

Ask three teammates about teams:

- What gets rewarded in our team?

- Who is the best team player on our team? Why—what do they do specifically?

- What is one of the stories in our team that tells us who we are and what we do?

TAKE ACTION

Consider the next likely challenging scenario that will confront your team.

- Who should you bring together to tackle the challenge to help them forge stronger bonds?

- How can you empower the team to take on the challenge in a way which creates a shared experience?

Decide on a 'one-percenter' or behaviour that you want to encourage in your team. Decide how you can reward and recognise it.

REFLECTIONS ON TEAMWORK

Teams work best when team members can play to their strengths and when the purpose behind the work of the team is clear. They are at the peak of their power when the values of the team members are aligned to the values of the organisation. Teams deliver exceptional outcomes when they work together on how they work together.

Supporting your team members to understand the work which lets them play to their strengths, and then allowing them to spend most of their time in this sweet spot, will see the individual and your team benefit. The future of work is not about completing the work we do now faster or better, but about doing completely different types of work. We know about IQ and EQ but the upcoming measure of success is likely to be AQ—the adaptability quotient. How quickly teams can adapt and change will determine how successful they will be and those teams with divergent strengths and skill sets have the best chance of being able to adapt.

> Uniting your team around values but being brave enough to give them the scope to think differently, work differently, and contribute differently, will be where Dauntless leaders enhance their team's ability to adapt.

I have joined teams where I was deliberately recruited for my different background but then forced to stay in the exact same lane as everyone else who did that role. I never fulfilled my potential in those roles and the leaders never got to take advantage of the difference I was bringing. They'd recruited me to do things differently, then forced me to do things the same. No-one won.

In other teams I worked with, Dauntless leaders put me into roles I didn't have the usual experience of background for, and they set me loose to do things differently. I still had to play by the same rules as my peers and was still bound by the same values, but the Dauntless leaders gave me much more scope to do things 'my way' within that framework.

I was able to complete tasks and solve problems in ways that let me play to my strengths, and so were my teammates. In these teams we won awards, created new products, improved customer service, made more sales, solved bigger problems and delivered more projects. The contrast was stark.

> When we all had to do things the same, everyone was stifled, but when we all had the scope to choose how we got things done, we were all better.

Leaders are often fearful they will be accused of treating people unfairly when they treat people differently. This stops them from allowing work to be done in different ways at different times from different locations. Dauntless leaders are confident in their own abilities to lead and work through ambiguity and subjectivity.

The thing is, treating people the same is not treating them fairly. Your most experienced person should have more freedom than your junior new starter. Your analytical thinker might benefit from more information and more time to consider options than your creative risk taker.

Treating your people differently is exactly what the best leaders I have worked with have done. They have treated us differently because they have tailored their style to each of us based on what we need.

This takes far more effort than treating everyone the same and requires a high degree of self-reflection and self-assessment to ensure you're providing an environment which is uniquely tailored to the individuals in your team but is still a fair environment.

> Shared values can help you treat people differently AND fairly.

When we all work under the same value set, then whether I get my work done from Brisbane or Sydney doesn't matter. When we agree to the standards that we deliver on and hold each other to, it doesn't matter if I get my part of the task completed at midnight or mid-morning, so long as it does not hold anyone else up.

If I solve the problem by talking about it aloud, consulting with others, reading and analysing, or synthesising data—as long as I solve it within the bounds of our team values, the method doesn't matter, it is the solve that counts. Spending my time doing analytical work while my teammate does conceptual work is neither more nor less valuable if both are required for our team to succeed.

Teams where the leader has made it clear that contribution to the team is what is most important also enjoy fewer ego issues and less self-centred behaviours.

Dauntless leaders acknowledge the behaviours they want more of and recognise people for how they help others achieve. They create opportunities for their team to collaborate and find communal success—to work together on projects and to address challenges and crises in a way that further unites the team. These become the stories that the team tell themselves about, who they are and what they do.

TEAMWORK—where to now....

REFLECT

Review your answers from the start of the teamwork section.

Consider:

- How much scope do you provide for your team to play to their strengths?

- What could you do differently about role design, flexible work arrangements and team structures to let people do more of the things they are good at?

- What do you see in your team around their shared experiences and values? What is observable? What do you wish was different?

GET CURIOUS

Ask three people in your team:

- What could we change about your work to let you do more of the things you're really great at?

- What type of connections do you have with your teammates?

- What story best describes who we are as a team?

TAKE ACTION

Talk to your team collectively about how you could design your roles and work differently to allow people to play to their strengths. How could you do this? Who would want to do what type of work? What would a fair distribution of tasks look like?

What are the projects, challenges or crises that your team could use to create or enhance your shared experiences? Use the next one deliberately in terms of who is assigned to the team and how they can use the experience to build stronger connections.

COURAGE

Dauntless leaders do what is right,
not what is easy.

We're not talking about courage in the traditional, masculine-focused sense of military bravery. Dauntless leaders show courage through the things they choose to do with the power that comes with their leadership role.

They are courageous in their approach to leadership and focus on how they can learn and **grow** rather than looking to prove how they are right. They are brave enough to admit their gaps, show their weaknesses and push themselves to the limits of their potential.

Dauntless leaders demonstrate courage when using their position or influence to **advocate** for others and help the people on their team to see their own potential. They create space for others to step forward, stand out and to achieve.

They show courage by being **authentically themselves—** the good, the bad and the ugly. They connect deeply with their team through the vulnerability of being exactly who they are.

COURAGE—before you read on...

REFLECT

CONSIDER:

- When did you last fail at something and how did it make you feel?

- When was a time that you achieved something you didn't think was possible? Why were your pushed or reaching for your limits?

- Who was the last person you advocated for? Why?

- What does authenticity mean to you?

GET CURIOUS

Ask three people in your team about courage:

- What does courage mean to them?

- How authentic are they at work about who they are, where they have weaknesses and what's important to them?

- When was a time they were at their best? What was it about the environment, themselves or the work that let them operate at that level?

TAKE ACTION

Record your reflections and the responses of others.

Read the chapters on Courage and use these notes to help you build a plan at the end of this section.

GROWTH
[THE STORY]

Dauntless leaders embrace risk and have a growth mindset.

The new guy who has joined our team is a hilarious bastard.

Pommy, pithy, and dry, within days of Kel's arrival he has become one of us despite the strange blue uniform he wears and the nuanced differences between Navy and Army practices where he declares things like "I'm off to hit the heads" when he goes to the toilet.

Kel forms up next to me as our troop stands on parade in the morning.

"Seventy-two Squadron. A-tttttten-shun!" roars the Squadron Sergeant Major (SSM).

Left knees raise and the slap of a hundred boots stomping to attention on asphalt echoes off the surrounding buildings.

The Officer Commanding takes over the parade and we're issued with the day's tasks before the SSM dismisses us.

"Faaaaaaall Out!" he orders.

We turn sharply to the right on heels and toes, come to attention again and as I stride out to take the two paces which finish the drill movement I slam straight into the back of Kel, who has remained rigidly at attention. The rest of my troop crash into me like a multi-car pile-up on a busy highway. Apparently, another nuanced difference between Navy and Army is that they don't take two steps after their 'fall out' drill command.

The Australian Navy, Airforce and Army all employ Electronic Warfare operators. EW operators are found on submarines and ships, in aircraft and vehicle patrols and within Special Forces and foot patrol teams. When posted to Canberra, Navy, Airforce, Army and civilians EW specialists all work side-by-side to provide a strategic collection of information. At several specialised bases around Australia, Navy, Airforce and Army EW operators work within their own service. There is just one exchange role for Navy within our single-service Army EW regiment, and Kel has been posted into that spot.

In late September 1999 I'm sitting next to Kel in our fanciest uniforms at a long formal table when we find our boozy dining-in night cut short. It's announced that we'll all need to be at work and sober the following morning with our trunks, stores, uniforms and equipment for the next six months, 72 EW squadron—gear, vehicles, EW assets and a hundred odd people—will all be deployed to East Timor within the next week. Our random Navy exchange bloke included.

"Bloody hell," says Kel. "I thought getting a Land Rover licence would be the highlight of my two-year 7 SIG posting."

〜〜

At the RAAF base in Darwin we sprawl out on the ground in the shade of an aircraft hangar. Our gear sits in a big pile beside us covered in red stickers in bold type declaring the trunks TOP SECRET.

"Can't believe that we were in such a massive hurry to get here and now we're just sitting around on the concrete waiting," I say.

"You've got to be kidding," says Kel as he props himself up on his elbow to face me. "That's like the whole deal of being in the Army—you guys do it all the time—hurry up and wait."

"Would've been good to have the time to tell Damo that I was leaving though," I mutter as I pick at the edges of a trunk sticker.

"Oh yeah, he's in Bougainville, right?"

"Been there for two months. I can't call him, he's got to call me and we usually only get to talk once a week. I haven't been able to tell him that the dogs are at our mate's house, the car is on the base and I'm on my way to East Timor."

Varush saunters over and sits down with us.

"I heard he only calls you once a week because you're boring as fuck," says V.

"I heard no-one ever calls you coz you fuck boring," I shoot back.

Jonno looks up from the book he is reading. "Who's a boring fuck?"

"V," I say earnestly as Varush shouts "Rach!" with conviction.

"What are you idiots talking about?" asks Chetty as he arrives with two jerries of water.

"How V is a shit root," I say with a laugh. Varush gives me a mock shove in protest as he gets up. He grabs Kel's water bottles along with his own and fills them from the jerry.

"Well, actually," says Kel, "before these two began an argument about who's the worst in bed, Rach was saying she hasn't been able to speak to Damo yet. He's deployed to Bougainville and doesn't even know she's here."

"I heard the head-shed talking about that," says Chetty. "They're trying to arrange a call but the satellite phones there are down at the moment. When you land in Dili they'll hook you up."

"Thanks, Chetty," I say. Chetty is our detachment commander and I know it's more likely that he went up to the HQ team to berate them about the situation and demand an answer, rather than the story he told us about simply overhearing a conversation. But he doesn't like us to know he gives a shit about us, so he regularly pretends it isn't him making our lives better.

V drops Kel's water bottles back to him full and hovers over me. I pull my two canteens out of my webbing and hand them to him.

"I also heard them say you, you bloody pusser," says Chetty using our term for Navy personnel and pointing at Kel, "that **you** are the hottest property in the whole landing force."

Kel shrugs. "I guess short guys with big heads are in demand."

"I think it's more likely to be about the fact you are the only qualified linguist in the first wave of landings, mate."

GROWTH
[THE LESSON]

When I joined the Army in 1997, all our training was based on the Vietnam War. Jungle warfare was the script for training and exercises, and our equipment was often still gear from the '70s. This included one radio referred to as a '77 set' based on the year it was brought into service.

It took me forever to reconcile the name of that kit with the year of its service introduction. I used 77 sets in **1997**, I was born in 1979, and all I could think was 'how could we still be using equipment from before I was born!'

As EW operators, during our initial training we learnt Morse code, we were taught how to touch type (because the assumption was no-one would know how) and we used huge A3 manuals to decode messages and research signals. Huge, bound, dot matrix printed pages with perforated edge filled with holes running down each side that you needed to carry with you to do the job.

In this same era, the internet was becoming more mainstream. Computers were becoming a household item, mobile phones were no longer carried in giant suitcases and emerging technologies were being adopted by consumers at an incredible rate.

I had landed in the communications industry at a time of immense change. The communications revolution—like the industrial revolution which came before it—was about to change the way we worked, the way we connected, and the way information was gathered, shared and held.

During the decade I was in the Army, the world went from snail mail and VHF communications to mobile phones, the internet, GPS tracking, smartphones, and drone surveillance.

To continue to do our job well we had to change to survive. A whole new set of skills would be needed in this new, digital world. But not everyone who did my job expected or embraced change. Not everyone had a growth mindset about being open to new ideas and an attitude of continuous learning.

There were a bunch of middle managers in my trade who decided 'what got me here is good enough to get me there' and so refused to attend any of the technical courses which would provide them with the skills to be able to understand this new technology. I mean you can't 'refuse' anything in the military, but people who don't want to do things can be pretty inventive about how they duck and weave out of things they want to avoid.

These same middle managers also avoided doing the longer language courses which had emerged as a vital element in the new communications era. Conversations, not Morse code, would be how opposition forces communicated. Long languages courses would make you unavailable for promotion courses. Many decided they'd prefer to go for the promotion.

These middle managers made an active choice. They saw the new courses (like the language course that Kel completed) and new technologies emerge, and they chose not to get involved. Their attitude was 'that's not for me' and they avoided the new opportunities to expand their capability set.

For some, I think they were afraid of the unknown and didn't want to risk trying something they couldn't be assured of their success at. They might not be good at learning languages and that could undermine their status. For others, I think they just wanted things to stay exactly as they were. A place where they were the established experts and were respected in the niche they'd carved out for themselves.

I think some actively resisted learning the new skills and showed disdain for others spending their time on emerging capabilities around languages, technologies or equipment, because maintaining the status quo benefits those already in the system.

> Those who hold the power in the 'old' system have the most to lose if things change.

Sometimes people don't realise that one of the reasons they might resist a change is because, subliminally, they understand it will devalue their existing skill set. If the world changes, their existing experience and expertise becomes less relevant, so often people cling to the 'old way' of doing things and in some cases, they try to undermine the 'new way'.

Whether they are conscious of what is driving this behaviour or not, in my experience those individuals whose response to change is to resist it, end up losing in the long term.

The middle managers who refused to learn the new technologies and take on the challenge of learning new languages soon found themselves lacking in the knowledge to do our EW job well. They couldn't make the right decisions in the new environment. They let themselves get so far behind the 'normal' skill set required for our job, they were no longer selected for promotion courses, nor for overseas deployments.

They were irrelevant, and that was of their own making.

Kel was one of the Dauntless leaders who chose to make decisions which pushed him in the other direction. He thrust himself into the new ways of doing things and continually found himself with an abundance of opportunities.

> Dauntless leaders embrace change and say yes to opportunities.

The year before we deployed to East Timor, Kel had spent twelve months at the School of Languages in Melbourne to learn Indonesian. It was a commitment to a full year of learning and a step out of a promotion pathway. He'd be out of contention for promotions, deployments, sea time and practicing the trade during that posting. He said yes anyway.

When we landed in Dili, the Territorial Indonesian Army (TNI) were still extracting. Being able to communicate clearly with TNI soldiers, local leaders, freedom fighters and militia in the volatile aftermath of the Independence vote and violence would be essential. Kel, with the language skills others had worked so hard to avoid acquiring, was pulled up and through the organisation to attend critical meetings with the highest-ranking people in the most potentially volatile situations. The random Navy bloke among an army of soldiers.

> Opting in, putting your hand up and pushing yourself to try new things and take on difficult challenges is something that Dauntless leaders consistently do.

Leadership is not about the rank you hold or power you yield with your position. The people who forge new paths in an organisation are leaders. The people who challenge the status quo are leaders. People who use their influence are leaders, along with those who stand up for what is right, who make change, who advocate for change, who get the work done, who do the crap jobs, who find the way through, who say 'yes', who say 'no'. We are all leaders when we choose to make a difference.

> The best teams are full of people with dauntless attitudes regardless of the rank they hold.

I believe the difference between Kel's attitude and the attitudes displayed by those change-resistant middle managers was mindset. Kel had a growth mindset and those other leaders had a fixed one.

Stanford psychologist Carol Dweck's work on growth mindset is a must. A growth mindset is one in which you have a tendency to believe you can grow, learn and develop, as opposed to a fixed mindset, where you believe your character, intelligence and creative ability are static givens we can't change in any meaningful way.

She argues—and her research proves—your mindset is something which is malleable and that you can change and cultivate it at any point in your life.

Dweck explains: "...growth mindset is based on the belief that your basic qualities are things you can cultivate through your efforts. Although people may differ in every which way in their initial talents and aptitudes, interests, or temperaments, everyone can change and grow through application and experience."

Those with a fixed mindset label people, ideas and situations in black and white; they are either right or wrong, a success or a failure. With a fixed mindset, people feel pressure to prove their credentials. How smart they are. How creative they are. How competent they are. It creates a constant need to defend their position, because outcomes are finite.

In contrast to this, a growth mindset creates the opportunity for people to identify their capability between these spectrums.

> With a growth mindset, a Dauntless leader.s focus is on what they can learn and how they could apply themselves to increase their capability in any area.

You can choose, cultivate and practice having a growth mindset.

Much of a growth mindset comes from how you frame a situation. Mistakes or falling short of a goal is an opportunity to learn, rather than a failure. Dauntless leaders don't worry about their image and acknowledge their imperfections.

> Dauntless leaders redefine failure.

People can see your imperfections anyway, so move from 'defending your position' to viewing your challenges as opportunities for growth. If you've built your team around shared values, your team will respect your vulnerability, your candour and your willingness to learn. You'll start to see the same attitudes coming from them when you frame your feedback using the same model, where the willingness to try something new is valued more greatly than only doing things you know you are good at.

Dauntless leaders are proud of their willingness to accept feedback and receive constructive criticism. They want to continue to learn, and they share this with their team. When people stop trying to project an image of perfection, they find many more opportunities to improve.

If you find it hard to face failure, consider a strategy that one Dauntless leader I know uses around 'yet'. What would happen if you added 'yet' to your self-talk about your success or capabilities? I'm not 'yet' good at tennis. I'm not 'yet' the number one sales star in my team. I'm not 'yet' an expert at this but I'm willing to keep trying. Framing loss, difficulties or failures in a way that allows you to learn from the past while looking forward to the future is a powerful way to redefine failure. This 'yet' approach gives you the chance to focus on what you have done so far, rather than getting fixated on the final part of the process and Dauntless leaders look at this in a different way.

Dauntless leaders value the effort over the outcome.

Telling people you admire their efforts rather than their outcomes builds a culture of growth in your teams. Benno telling me he admired my determination to keep running is more valuable to my growth and attitude towards trying something I struggle with, than telling me I'm good at running.

If someone is encouraged because they've worked hard to improve in an area, they will be more willing to continue to put in the effort to develop or to apply themselves in learning something new in the future.

When we reward our people for being smart rather than for their work to solve the problem, then they'll want to maintain that image of being smart and may avoid trying new things.

Someone with a growth mindset and one which focuses on their efforts will be more willing to continue to work at something they don't succeed at right away and to take risks on trying new things.

Dauntless leaders value hard work over inherent skill.

My favourite quote is this Churchill gem: 'Success is not final. Failure is not fatal, it is the courage to continue that counts.' It values the struggle, the determination to persist and the resilient attitude that it takes to learn something new that you're not immediately good at.

Kel not only became a key asset in East Timor, but his continued attitude to take on new challenges and try new things opened the door for all sorts of unique experiences. He became a submariner, was assigned to covert operations, studied for a Masters degree, and became an officer. He posted overseas, became an emerging technologies specialist and studied for a second Masters degree. His mindset was that he could learn, he could develop, and he could grow. Being that he started in the Navy as someone who barely who skated through his final years of school, he showed me the right mindset can open all sorts of doors if you are prepared to work hard to get there and to take some risks.

Dauntless leaders take risks and face challenges with grit.

They take on projects with opportunities to grow their leadership capability and often find their best growth in those areas which are most challenging, and even where they fail.

Sustaining your effort and continuing to try in the most challenging environments is where you find out what you are made of. You learn very little about yourself and others when times are easy, it is when times are tough you see people's true colours.

You see who puts the team above themselves, who is willing to keep working even when things are tough, who finds a way to have a positive attitude even when it seems impossible.

Stop worrying about how a failure will look to others and focus on what you can learn in a difficult environment.

GROWTH
[THE APPLICATION]

Over my career I have had many inspirational leaders. Men and women with open minds, growth mindsets and a willingness to challenge the status quo. Dauntless leaders who also encouraged those in their charge to grow.

I will never forget the focus on growth from the senior women I met early in my military career. While they were breaking down barriers and having to deflect every comment that suggested the opportunities they earned were only given because they were female, they still found the time to encourage the young women working in their teams to stretch themselves.

These women had faced enormous obstacles at every step of their careers and had dauntlessly persisted. They did everything they could to make sure those obstacles did not exist for us.

During quiet conversations, hard-faced, stoic leaders who had tried to scrub away all their femininity to force their inclusion, shared vulnerable, open stories of their own struggles and unrecognised triumphs.

I was lucky to come after these Dauntless leaders and enjoyed the benefits of the paths they forged to open roles in the Army that were previously closed to women.

One of my earliest military leaders was Mac, an inspirational woman who was a very senior leader in our team. She took me aside one day and gave me advice that has stayed with me and helped steer me on every twist and turn of my career:

- Say 'yes' when presented with opportunities.

- Do the things you think you can't.

These two attitudes helped me to consistently grow. It was great advice for my time in the Army and was also great advice for the rest of my career and life. Having another woman help me understand, so early in my career, how I could forge my own path, was instrumental to my development. I could 'see it'—she had got there—so I could 'be it'.

It helped me have a mindset where I was open to taking risks and less worried about failure. With these two attitudes— saying 'yes' and doing what I thought I couldn't—I was ready to learn and grow at each opportunity.

> I was open to risk, open to failure, open to trying things I wasn't sure I'd succeed at, and because I had a growth mindset.

Off the back of this attitude I learned that when I thought I was at the limit of my abilities, I wasn't.

The Army teaches you how to access your own massive, often untapped well of potential by pushing at your every boundary. When you think you've gone as fast as you can, they ask for faster. When you think you've climbed as high as you can, they ask for higher. Distance, pace, accuracy, decision making, mental endurance—in every sphere they test you, push you, and ask for more.

You learn how to find the edges of what you think you are capable of, the point at which you think you have nothing left to give and you learn that this limit is in your mind. You **can** push at those edges and move them outwards to be faster, be stronger and do more.

We can all do so much more than we think we can. We can dig deeper, go faster, do more, learn more, achieve more, keep persisting and make a mark on the world far beyond what we think we can. We all have this within us; both the mental limitation on what we think is our 'edge' and the ability to push past it either through intrinsic or extrinsic motivations.

> Once you know you can stretch the edges of your limits outwards, anything seems possible.

It's an enormous gift to learn this about yourself. It's a perspective which stays with you for the rest of your life and it's one many people don't even get to glimpse. It helps you to become a Dauntless leader yourself, and wires you for a growth mindset.

In the army these challenges were mostly physical or placing myself into unusual situations. I said 'yes' to a machine gun course and to working in Suai with 1 RNZIR, then 2RAR and the Occussi Enclave with 3RAR. I pursued things I thought would be impossible, like learning new languages, training for a parachute course and running a half marathon.

In Afghanistan, I did things I never thought I'd get the chance to that were new, difficult and I wasn't sure I'd succeed at. Working from armoured vehicles meant finding new ways to deploy our EW gear. Embedding with combat troops meant I needed to physically perform to keep up. Being empowered to make the decisions on the ground meant I needed to get comfortable with really big calls that leaned heavily on my language and analytical skills.

During my time in the Army, I also got to fly a plane, shoot a sniper rifle, drive an ASLAV, and cut donuts on the beach during my Unimog truck driver's course. All super cool things that you definitely need proper qualifications for, but because I was there and because I was part of the team someone said 'want a go?'—so I said 'hell, yes'.

> Holding a growth mindset works outside of the military world too.

I said yes to a Bank Manager job I was very uncertain about.

I'd only been discharged from the Army for a few months before I decided banking would be my next career. I'd done up my resume and walked around town dropping it off to local branches. I was hoping to secure an entry level job and was excited about starting a new career.

One of the places I dropped my resume had a Branch Manager role vacant. The Regional Manager decided they'd had enough failed experiences in promoting bankers into leadership roles, so decided to take a risk and appoint someone with leadership skills, but who didn't know banking. I was that risk for them, but saying yes was also a big risk for me. It was a risk which helped me forge a whole new career.

After seven years working in the retail network where I led branch teams, lending teams and regions and I was responsible for a balance sheet of over $3.2 billion dollars, I decided I was ready for a new challenge and to take on a new risk. I asked my boss for his help.

I'd enjoyed leading projects working with teams, things like helping the bank figure out how to manage their cash handling differently, or running pilot programs for different team structures or processes, and I wanted to pursue this further. I wanted to see if I was capable not just of leading projects in branch teams but for programs which would impact people across the whole of the bank. There were 40,000 people in the organisation with only 11,000 of these working in the branch network. I wanted to start doing things that would change the working lives of our whole organisation.

I took a risk and told my boss the truth. I basically said, "I don't want to work in your team anymore." Being a Dauntless leader, his response was "no worries, what do you want to do next? Let's figure out how you can get there."

Find yourself Dauntless leaders to work for and with.

Dauntless leaders—like my boss at that time—are not afraid to let people go on to new things. They don't try to keep people in their own team because it's better for their own scorecard. They focus on the individual and support them to learn, grow and develop whatever team they are in.

I'd highly recommend working for awesome people like this where you can. Your boss will have a massive influence on the opportunities you get presented with and your ability to do new things. People who insist on holding you back for their own benefit will always put themselves above their team. Leave those leaders and find a Dauntless one.

I was hoping for a sexy, big-budget, exciting, transformational project, but of course I hadn't earned the right to do anything like that just yet. Instead, I was offered a three-month secondment to close down a division in Singapore, managing the customer experience, the staff redeployment and redundancies, and to complete the whole thing remotely from Brisbane with no budget.

I'd never worked with that part of the business, or in Singapore. I'd never worked in projects and I'd never worked for the brand this project came under. I was absolutely sure that I would find lots of things I didn't know how to do, but I knew I would learn a lot from the experience, so I said yes. It didn't look anything like the sexy big-budget opportunity I was hoping for, but I said yes anyway.

> The opportunities you want don't always start out looking like you think they should, but they often become the thing that opens the right door.

That piece of advice I received from Mac way back when I was nineteen helped me be much more Dauntless in my attitude towards risk and opportunities. To just say 'yes'.

Say 'yes' to the secondment that isn't quite what you'd described to your boss. Say 'yes' to the project that seems like it's a time-waster. Say 'yes' to the function, to the networking, to the course, to the training, to the promotion. The worst thing that'll happen is you'll learn what you don't like, and even if that's the outcome, you'll still have benefited from the experience.

That 'not quite what I was looking for' project opened the door to a role in Inclusion & Diversity where I was given the opportunity to lead the development and implementation of a truly unique, innovative and groundbreaking approach to Autism Hiring. While in the Inclusion & Diversity team, I also led programs to increase flexible working, to support women's leadership development, and to create inclusive leadership learning programs for leaders at every level of the bank. Collectively, across these streams of work I was helping to shape a better workplace for 40,000 people. The 'not so hot' job opened the door to the 'job of my dreams'.

When you start to do the things you think you can't, you get more comfortable taking risks and facing challenges, but you can still find yourself doubting what you can achieve.

I was terrified when preparing for my TED talk on Commander's Intent.

Not for the normal 'fear of public speaking' reasons, but because I was worried about being crucified by my military peers for talking about my time in the Army.

I felt uncomfortable with the editing process which meant that to tell a compelling story in a way which was concise and meaningful I'd need to omit key details or qualifying sentences.

I was worried other veterans would be quick to look for any chink in the armour, or for those who hadn't worked with me to question the validity of my experiences, or even opinions, about what I had learnt.

The manuals on military leadership and Commander's Intent are hundreds of pages long. I only had twelve minutes to explain it, simplify it and relate it to leadership outside of the military and the future of work.

The military community can be harsh. They love to troll anyone who makes a claim which can't be validated. They love to cut apart anyone whose experience might be different from their own. I was so worried I'd be ridiculed by the people

I was attempting to validate, that it almost stopped me from doing it. Almost.

Instead of backing out, I listened to those around me telling me that I could, and I stuck to Mac's advice and did the thing I was afraid to do. I'm glad I didn't back out and I'm proud to say my experience was the opposite of what I was worried about.

I found my old teammates supportive and encouraging. Peers from across Navy, Army and Airforce reached out to tell me they had my back. People I'd never worked with sent me messages to encourage me to continue to talk about my service. Veterans could see their own successes, and feel the value of their own contributions reflected in the stories I told.

I'm sure there are haters, and trolls shit-talking me, but I am lucky to have a fierce community of people from the military, and from the rest of my life, who have been so amazing and supportive. I'm glad I did it despite my worry that I may fail, make a mistake or be ridiculed for trying.

> Having a growth mindset about success, failure, and persistence gives you the chance to take on more risks and learn about your true capability.

Focusing on your effort over the outcome encourages you to say yes to the opportunities which come your way. Having the attitude you should take on things you're not sure you will succeed at creates these opportunities to learn about yourself and the people around you. You'll test where your edges are, and by jumping into these roles and situations you get a chance to learn what you are really capable of.

ACTIONS—GROWTH

REFLECT

Reflect on the last time you failed at something or made a mistake.

- What was your self-talk at the time?

- Were you focused on the effort or the outcome?

- What did you learn that helped you in situations after that error?

- How comfortable do you think you are with taking risks and saying yes to things you're not sure you will succeed at?

GET CURIOUS

Ask three teammates:

- What's the last risk they took? Why?

- What did they learn about themselves when they took the risk?

- What stops them from saying 'yes'?

TAKE ACTION

What's something you've been avoiding doing/learning/ trying? Sign up for it! What's the worst that can happen?

Ask the next three people you talk to for some honest feedback about something you could be doing better. Practice receiving this feedback as something that you can work on and grow from rather than a failure or a weakness.

Who on your team could benefit from understanding more about a growth rather than fixed mindset? Book a chat with them this week.

ADVOCACY
[THE STORY]

Dauntless leaders back their team and lift others up.

"I'll take Pavo," says Fred.

Pavo grabs his weapon and glances at the people beside him. He walks to Fred, turns, and faces back towards the group.

Two other Corporals stand beside Fred, and he lifts his eyebrows at them. The taller one sports dark wavy hair at a length which is pushing the boundaries of a compliant military haircut. His shirt stretches tight across his broad chest as he lifts his hand to take a long drag on his cigarette. Fred gestures to him. "You're next, Blocky."

Blocky exhales, scratches at his neck and looks out over the soldiers gathered before him. They'll each need to choose nine soldiers to join their team, and the most important one is who they'll choose to be their Second-in-Command, their 2iC. With Pavo already selected as Fred's 2iC, Blocky makes a quick decision.

"Jess," he says and beckons her with a wave.

She joins him with a grin and they exchange a few hushed words to begin planning their next pick.

The two most senior soldiers are now out of contention.

Jess and Pavo have both completed their promotion courses. The rest of us are fairly new. Some of the group assembled have only graduated from their initial training a few months ago, and this will be their first exercise. Others, like me, have been at the regiment a year or two, but are still junior soldiers. The final section commander, Al, will have to choose someone without rank to be his 2IC.

Al's eyes take a measured tour of the group and he lingers on a few faces as he thinks about his options.

"Walks," he says after a considered pause. "So long as you promise not to be a dickhead."

"I promise nothing," Walks says as he swaggers forward to join Al.

I shuffle on the spot as the section commanders continue to choose their teams for the upcoming Infantry Minor Tactics (IMTs) exercise. It will be three weeks of patrolling around 10 kms a day at the High Range Exercise Area in Townsville, Northern Queensland. It will be hot, dry, hilly, and hard.

In late 1998, the only deployment on offer for EW operators is a peacekeeping mission to Bougainville. Big exercises are the peak of Army activity in this pre-Timor period and IMTs is an annual major exercise for 7th Signals Regiment.

American forces are here to play enemy, and we'll be rehearsing patrolling, securing landing zones for helicopter touchdowns and section attacks. We'll be carrying 30kgs of gear each for the duration. It will be physical.

Those who are the strongest and fittest are sure to be chosen first and I figure I'll be picked very late in the draw. I take a look at a few of my mates and see the same self-doubt in their faces. We know that with every selection the section commanders are considering how they think you'll perform when faced with the physicality of man-packing your gear for weeks and weeks.

Everyone already knows who is fit, who is strong, and who is not. Your strengths and weaknesses have been discussed, considered and on display in the lead up to the exercise. Your capabilities are known to everyone.

I think about what I bring to the team. I'm terrible at running, have just completed my Minimi machine gunners course, and am decent at EW. Unfortunately, EW is not part of Infantry Minor Tactics.

"Okay, Machine Gunners next," says the Sergeant in charge.

"Fred, you pick first. Everyone with a Minimi qualification step forward."

I step out. I'd been hiding up the back and now am forced to stand and face a public selection process. There are six of us who are machine gun qualified. To make up the three sections they'll need all of us—two machine gunners per section—it's just a matter of who will join which team.

You want your machine gunners to be the biggest, most burly dude-bro blokes you can find. The weapon itself is a few kilos heavier than a regular rifle and carrying the link ammunition puts you at about double the ammo weight of other section members.

Machine gunners also need to be aggressive. They need to storm up hills to hold the high ground, to fight with ferocity to occupy a position. They are the biggest firepower in a section attack, and they need to be up the front and pushing for ascendancy.

I look left and right at those who've stepped forward with me. Five big, tall, strong guys, and me, 55 kgs wringing wet.

Fred lets his eyes rove over us. Pavo murmurs his opinions in Fred's ear.

I look down at my boots and wait for the humiliation that is sure to come when someone is forced to take me in their team as the last option.

"I'll take Rach," says Fred.

And with that decision he changes the whole shape of my military career and everything I think about myself.

〰

On a steep hill climb on the fifth day, I trudge towards the top.

I look out for a few beats and scan for movement as we climb slowly towards the peak. I make eye contact with Pavo ahead of me in the formation and, as I keep moving forward, I step in a circle to look backwards to connect with my teammate behind me. I get a chin lift from Pavo and a head nod from the bloke behind me. No-one makes any hand signals, no-one has any messages to pass.

We have about three kilometres to go for the day, which will take us up and over this hill and skirting to the west below the ridgeline before we come into the night location. We have already stopped for a resupply, so our water bottles are full, our ammunition pouches are bulging, and we have another three days food jammed in our packs.

It's late afternoon and the sun feels like a furnace. We doggedly walk towards its bright orange glow and I know that as we crest the peak we will spend the next hour staring into it as we track west. The scrubby, spindly trees offer little shade but many obstacles. My shoulders ache from the weight of my webbing and pack and I step short to shuffle and skip, tugging on the pack straps to cinch the weight closer to my body.

Fred works his way towards me.

"How're you doing, Rach?" he asks, low and quiet.

"I don't know if I can do this," I mumble.

"Yes, you can. Keep going," he says as he moves back to the centre of the formation.

〰

The next morning we load up with all our gear and patrol for a few hours before we hear the pops of the American forces firing at us with blank rounds.

"Contact front," yells Pavo as he flings himself to the ground and returns fire.

We all sprawl to the ground and my pack smacks me in the back of the head as I dive behind a tree.

"Eleven o'clock," shouts Pavo, indicating where the enemy fire is coming from as we all start to return fire and bound into positions.

I dump my pack as I hear Fred yell, "Get that gun up the hill. I want the high ground, Rach."

"GUNS GO!" he bellows.

"GUNS GO!" we echo in unison, shouting at the top of our lungs while we get to our feet, run ahead for half a dozen steps before flinging ourselves to the ground again and continuing to return fire.

Working with the two riflemen in my gun group we yell, bound and drop a dozen times to get up to the higher ground on the flank. It's effectively running up a rocky hill doing burpees every few steps with piles of gear on while carrying a log. Sticks and rocks stab at me as I work my way up the hill and, once in position, we start to fire bulk blank rounds so the rest of the section can push towards the enemy position.

We stop shooting as the rest of the team crosses in front of us and we continue to shout and echo the commands for the rest of the team. Once we hear "three dead enemy" shouted from Fred, we get ready to move again for the re-org.

I sit on the ground with my back leaning up against a tree while we reorganise and account for rounds, rations and water.

The gun sits on its bipod beside me on the dusty ground as I open pouches and count ammo. I gulp down water as Fred checks in with each of us individually. I'm still breathing heavily by the time he makes it to me. The adrenaline has worn off and I'm feeling shattered.

"How're you doing, Rach?" he asks.

"I don't know if I can do this," I say.

"Yes, you can. Keep going," he says as he moves to the next person on the team.

〰️

"Okay—these are my orders, no questions until the end," says Fred as we hunker down around a crudely drawn mud map of our day's patrol route. He works through the current situation, our mission for the day, enemy information and expected interaction, admin, logistics and comms for the patrol. Not much has changed since yesterday; another long patrol, expect enemy engagement.

There is a fifteen minute notice-to-move window so we head back to our positions in the harbour. Our three sections have all come together at the end of the second week and I share a pit with Pavo.

I pull out a wedge of waxy hexamine and light it to get some water boiling for us on the tiny stove. Pavo checks on the rest of the section before coming back to our pit. He strips off a boot and a sock.

"Argh, my bloody feet."

I glance at him and he waves his heel in my direction. A huge swathe of skin is pushed up behind his ankle. The exposed skin below is raw and red. Weeping and angry.

"Gross," I say in agreement.

I look out to the front of our position and maintain our piquet while he tapes up his foot in preparation for today's patrol and starts talking about breakfast. I listen to him verbally run though his dining options. They all sound shit. I am sick of eating ration packs.

As we push through into the third week, long gone are the days of rapidly consuming every last skerrick of the cans of beef casserole, ravioli pasta and random bits and pieces that make up a one-man ration pack. My appetite is zero and it is the same with most of my teammates.

"How's everyone going?" I ask as Pavo works on his other foot.

"Danny's back is bad, had to get Tozza to take the other gun for us."

"Want me to go and take her through some stuff?"

"Nah, it's okay," Pavo says as he drops into the pit to take over the piquet. "Danny has been showing her the stoppage drills, they'll work out the rest."

A battered and crumpled packet of M&Ms seems like my most appealing breakfast option so I tear it open with my teeth and pour the chocolatey-goodness directly from the packet into my mouth. The sweet taste is ruined by the whiff of petrol that lingers on my fingers from the hexamine.

Around my mouthful of chocolate I garble. "How about Ems, he looked pretty crook yesterday?"

"That bald bastard just didn't drink enough water," says Pavo. "What an idiot. He should know better. I expect that from the jubes straight out of training but not from him."

I pull apart my weapon, cleaning and oiling the working parts. I'm nervous that I'm only just holding it together but I want to help the section. "Want me to take the extra link he was carrying for Tozza's gun?"

Pavo pauses for a second. I see him mentally adding the extra kilos to my pack. I'm already carrying 200 rounds on the gun, another 200 in my webbing, and 400 extra in my pack. "If you can?" he says.

"Will it leave them short on rounds?" I ask as I quickly put the gun back together. I've managed to guzzle mouthfuls of chocolate and a muesli bar as I go. The Army teaches you to shovel your food in and chew with your stomach later.

"They'll be right, if we get in a contact we can bomb them back up afterwards."

I slap down the feed cover and complete the final function test feeling the bolt slide forward and slam into the end of the empty chamber. I reopen the tray and slide on a belt of blank ammunition as the water starts bubbling on the tiny stove beside me.

"I can do it," I say, not sure I really can.

"Brew?" I ask.

I siphon off half for Pavo to shave with and with the rest I make us a coffee so thick with condensed milk the spoon can almost stand up in it. We take turns blowing steam off the top of the brew and trying to avoid touching the lava hot metal canteen as we share it between us.

A medic comes around to check on each of us. There are a few people who've already been pulled out of the exercise with knee and ankle issues. Others nurse niggles from old injuries or new ones.

"How's it going, you two?" he asks me and Pavo.

"Good, mate," says Pavo. "You know about Danny. Ems looks better today."

The medic throws Pavo some extra blister dressings and looks towards me.

"Yep, I'm okay."

The medic moves off to the next pit and Pavo gets up to collect the extra belts and magazines of ammo from Ems so I can load up my pack before we leave.

"First positive answer to that question I've heard all ex," he says as he adjusts his webbing and heads off.

〜〜〜

ADVOCACY
[THE LESSON]

Dauntless leaders lift others up.

Sometimes they're actively choosing to take a stand on an issue and sometimes they are supporting an individual they believe in, regardless of their differences or challenges. But there is a pattern. Dauntless leaders advocate for others, lifting their team members up and making space for all their people to succeed.

I survived that exercise because every time in those first few days and weeks where I got to the point where I thought I couldn't do it, Fred would be there telling me I could.

By the third week, I knew I could do it, too.

My head was up on patrols, I marched the gun up the hills, poured rounds down on the enemy positions, occupied the highest ground, started looking for better locations for us to gain ascendancy. I asked Fred for more tasks. I carried extra weight.

When we got to the end of the three-week exercise, I was still there, still carrying the gun. Out of the six machine gunners picked from the original line-up, I was the only one still lugging the gun at the end of the exercise. And everything changed.

Suddenly, my reputation was that I was tough.

I'd overhear others around the barracks saying things like, "Oh yeah, she can't run for shit, but she can carry a gun like a motherfucker."

Both true.

I started to believe it too. With my new-found self-belief and a reputation which grew with every exercise, I said yes to any opportunity which came my way.

Want to go to East Timor? Yes.

Want to embed with the New Zealand Infantry? Yes.

How about supporting Australian Infantry in the Occussi enclave? Yes.

You should train for a parachute course. Yes.

We want you to be the analyst on a high-profile desk when you arrive in Canberra. Yes.

Language course? Yes.

Promotion? Yes.

Wanna carry the gun again? Yes.

Can you figure out how we deploy our gear better? Yes.

Afghanistan? Yes.

You should be on the first patrol. Yes.

Embed with Armoured? Yes.

How about Australian Infantry foot and vehicle mount embed? Yes.

No wonder our household default option is 'yes'.

Mac's words rang in my ears whenever I was offered a new opportunity. Her encouragement to keep saying yes to the things I thought I couldn't do. I heard Fred's voice too.

Supporting me when I thought I'd gone too far and reminding me I could get there if I just kept working.

You never know the moments that are going to define you and shape your life. Carrying the gun on that exercise, defined me and shaped my life. Any chance you get to do something you don't know if you'll succeed at—grab it with both hands—you'll never learn more about yourself.

Once I started believing I could do it too, wherever leaders wanted to stretch the boundaries of what women were allowed to do in the Army, I was given the chance to step forward. It was awesome. Fred's decision to choose me on that exercise had changed everything about what I believed it was possible to achieve.

〰〰

I went to a reunion not long ago for the regiment. One of the Warrant Officers who'd been on that exercise with Fred was still kicking around the regiment and came up to share a beer and conversation later in the night.

He introduced me to one of the younger blokes who arrived at the regiment long after I had discharged.

"This is Rach," he said. "She can do anything blokes can do."

That exercise was twenty years ago. And I can't do everything a bloke can when it comes to being in the Army; I probably never could.

But the reputation I forged off the back of Fred's decision to choose me, support me and encourage me to succeed on that one exercise lasted the rest of my military career.

As a junior soldier I was selected to brief the then Chief of Army, Major General Cosgrove, when he came to visit the regiment. I was selected on flight missions, given special assignments and unique opportunities, and when it came time to go to Afghanistan, I was embedded with frontline combat patrols almost every week, because my reputation was that I could do it.

My reputation went from wimpy and bad at running to the woman we put to the front. The shape of my whole military career was changed because one person, one time, believed in me.

> At a time when I didn't believe in myself, a Dauntless leader showed up and told me I should.

We often don't know the long-lasting impact we make on others. A feedback session which gives someone an insight into themselves, a casual compliment someone remembers every time they think of you but long after you've forgotten you said it. Those single moments, decisions and conversations can cause a ripple of change we may never even see.

> Every day we have the power to impact people and influence how they feel about themselves.

I didn't ever get a chance to ask Fred why he chose me as his machine gunner for that exercise.

Was he trying to challenge the status quo on gender?

Was it a strategic move I never knew the reason behind?

Did he just feel sorry for me?

Regardless of why, he made a difference when he picked me, but he really changed the game when he followed through by supporting me. In the first week on that exercise there were times when I was barely holding back sobs as we patrolled because I just didn't think I could do it. It wasn't going to be a matter of if I would fail, it would be when.

Fred never sat me down to deliver an impassioned speech. He didn't pepper me with inspirational quotes or try to motivate me with guilt. He was just there at my side—at all the times when I was at my lowest—telling me I could do it.

His belief in me forced me through my limits.

At many points in those first two weeks I wanted to pull out. I thought I had given my all and I had nothing left, but I was wrong. Over the next decade the Army taught me over and over again; when I thought I was at my max, I was still nowhere near it. Turns out, there was plenty left, I just needed to find a way to dig deeper.

On that exercise Fred's quiet support helped me to push through and re-establish new limits. When I thought today was hard, I remembered what I had achieved yesterday. So I kept going. I needed Fred less. I started to believe that I could make it.

Fred was responsible for nine of us. Nine junior soldiers. All relying on him to navigate, make decisions, get the gear where we need to go, manage our wellbeing, manage our egos, triage injuries, and achieve results. Still, he found the time to notice us as individuals and advocate for us to be given new opportunities and to be supported to succeed. Dauntless leadership.

He knew when we looked good, he looked good.

But he also just gave a shit about us.

ADVOCACY
[THE APPLICATION]

We all have the power to make this happen for others.

We can all support someone, show them we believe in them, and back them when they don't back themselves.

This is leadership.

Helping people to stretch, grow, and challenge themselves is one of the true pleasures of being a leader. You see the light in their eyes go on when they realise what they are capable of. And you see them start to choose to do the difficult things and you delight in their confidence and achievements.

You don't need to be someone's boss, or even hold a formal leadership position.

> Whether we lift others up or not is a choice we get to make in almost every interaction we have.

I've seen my son lifted by a few words from a coach. A couple of comments about his courage in defence during the half-time break reverberating through years, and years, and years of his football playing career. The coach noticed his tenacity and called it out in front of his peers; my son's self-talk, then his image became "I deliver tenacious defence."

Sometimes lifting others up is noticing what someone has done and recognising or amplifying it. Other times, it's encouraging them, telling them you believe in them before they even start.

Being someone's cheerleader can be loud and showy or just making a choice to pick an underdog. Your words matter. They echo and resonate in ways you may not always consider. Using this power for good, you can help individuals and change the shape of the world around you.

> People live *up* to your expectations.

If you've acknowledged their bravery, they want to be brave. If you've told them you know they can succeed at a challenge, they want to succeed.

When you encourage, lift, and support people by telling them you're proud of them and you believe in them, you develop yourself as well as them.

Lifting people up is not just for your star performers. As one insightful teammate told me when we were discussing her year: No-one comes to work to do a shit job.

And in almost all cases, I've found this to be true.

It can be so easy to get caught up creating mountains of paperwork, policies and rules for our organisations in an attempt to pre-empt or prevent people 'taking advantage of the rules'. But all we end up doing is making it hard for people who were probably going to do the right thing anyway.

One of the things leaders don't always recognise is when they lift others up, they are using their influence for impact.

Who they choose to lift can help them build a more inclusive and diverse team.

When we make choices to amplify the voices of those who struggle to be heard, or to encourage someone who might have felt they were on the outer to be included at the table, we impact the individual, our team and we also start to shape the world.

Whether Fred wanted to bust gender stereotypes or not, by choosing a non-conventional candidate he made me feel more valued and included, and his decision had a ripple effect on our section and other teams, changing the way others thought about women machine gunners for years after that initial decision.

Ripple effects can be far reaching. You may never see the wave of success crashing on a distant shore because of the pebble of encouragement you threw years before.

Knowing this, you can *and should* be deliberate about who you lift up.

Who do you mentor or are you a mentee too?

Both relationships reveal an insight into the type of person you respect and admire.

Do they look like you?

Think like you?

Have the same background as you?

Come from the same socio-demographic group as you?

When your networks are more diverse, you stretch your world view, but also you stretch the impact you can make.

Who do you promote?

Lifting people up is not just about the words you use. It is also about who you chose to promote and who you give challenging opportunities to. Think about the last three people you promoted. What ages were they? What gender were they? How different were they from each other?

When you think about promoting someone or bringing a new member into your team, do you focus on industry experience (so writing things like 'must have five years' experience' on your job ads)?

When you do this, you exclude people from other industries who will bring you new insights, people who are changing careers, veterans, neurodiverse thinkers. You are asking for the 'same'. Is that what you really need—more of the same?

Have a look at your customer demographics. Who are your customers? Where are they, what are their backgrounds, how do they culturally identify, what are their sexualities, abilities, genders?

Your team should reflect the customers they serve. Do they?

Who do you bring into your circle?

Look around the room at your next committee meeting, volunteering night or study group.

Is there a wide range of people there—different ages, backgrounds, sexuality, cultures, and beliefs? If not, how can you invite someone into the circle who would help you to gain a different perspective, and would help them to broaden their network?

Reflecting on who we mentor, who we promote, and who's in our inner circle can shine an uncomfortable light on our own unconscious bias when it comes to showing up our preferences for 'I like people like me'.

Dauntless leaders advocate for others who are like them **and** those who are different to them. They acknowledge the power they hold as an individual and they use this power to make the space for others to succeed.

When you start advocating for others, you'll see the enormous impact it can make on individuals, on groups and on your organisation. As one Dauntless leader whose passion for advocacy was evident across their team said to me:

> The thing I love most about being a leader is
> the opportunity to play a small part in helping
> others succeed. I may have only opened one
> small door, given them one small push or

advocated for them with one small sentence
and then—BOOM—they are off and running.
I rub my hands in glee at them going on to
succeed at all sorts of things that have nothing
to do with that one small action at the start.

Opportunities to advocate for others come from all sorts of places.

When I was awarded Employee of the Year in the Prime Minister's Veterans Employment Awards in 2018, I realised I could leverage the award and use it as a platform to advocate for Veteran's Employment.

For the twelve months I held the award I would speak at the opening of a letter if I thought it would provide either the chance to show employers that veterans could be great employees for their business, or to speak with veterans about the transferable skills they could bring from Defence into their next career.

The two groups (employers and veterans) often overlook the skill set veterans discharge with. What you bring with you when you leave the ADF goes well beyond firing weapons or working with specialist military gear in specialist military circumstances.

In a 2018 data analysis, LinkedIn identified the top twelve skills Australian leaders look for in employees. Those skills are:

1. Management.

2. Leadership.

3. Project Management.

4. Change Management.

5. Strategy.

6. Strategic Planning.

7. Program Management.

8. Business Strategy.

9. Negotiation.

10. Business Process Improvement.

11. Business Analysis.

12. Business Process Management.

Veterans over index on *all* identified skills but of particular strength are Program Management (six times the national average), Change Management (three times the national average) and Leadership (two times the national average).

Although they may lack technical, industry-specific experience, veterans bring exceptional abilities to teams. Their different approach to problems, their perspective, and their ability to push through to a solution are some of the ways I've seen veterans help their teams to deliver on projects and outcomes that were stuck or stagnant.

It will take employers being more open to the skills veterans can bring and veterans being more descriptive about how they've used these skills to close the current gap between veteran unemployment, and the general unemployment numbers in Australia.

ACTIONS - ADVOCACY

REFLECT

Who have you mentored, advocated for, or promoted recently?

How different/similar are they to you when you consider gender, age, cultural affinity, identity or background?

Consider how you use your power or influence now to advocate for others.

- What is it that you do?

- How deliberate are you in these actions?

- Anything you'd like to do more of?

What are the causes or issues you care about? Can advocacy help you spend more time in these areas or to help you identify how you can make a difference to these causes?

GET CURIOUS

Ask three people:
- What does advocacy mean to you?

- Who has advocated for you in your career or life?

- What impact did that advocacy have on you (long and short term)?

Ask someone you've advocated for in the past about your words and actions. Ask if they made a difference to them and how.

TAKE ACTION

Find someone you can lift up today.

It might be someone you can advocate for who's being considered for a project or promotion.

You might be considering who you next mentor—how can you make that a deliberate choice to change the game for someone?

Be conscious of your interactions throughout the day. Are there moments where you can provide encouragement or help someone to believe in themselves.

AUTHENTICITY
[THE STORY]

Dauntless leaders are authentically themselves—the good, the bad and the ugly.

Our interpreter walks ahead of me holding a communications receiver in his hand. A slim black cord traces a thin dark line from the device, up and across the sandy tans and blondes of his desert-camouflaged chest and into his left ear.

I look up towards the front of the patrol and watch the swivelling heads and the smooth deliberate gaits of the infantry soldiers we are embedded with. Their elbows protrude wide from their bodies as they nurse their weapons across their chests, each with his own style. Although issued the same Australian desert-camo uniforms as the rest of us, Bahiri stands out like dog's balls with his hunched shoulders, overweight body and shuffling feet. He bravely keeps the pace and checks over his shoulder to make sure he is close to me. He goes back to watching the footsteps of the rifleman he's been told to follow while he listens.

We parallel a wide ditch alongside a mud-walled compound. Poppies grow tall in the fields across the ditch. I keep one hand on my weapon, finger outside the trigger guard and reach up to touch a fingertip to the hard plastic of my own earpiece as it squelches in my ear. I squint. I always find squinting helps to listen harder. I listen for a few seconds and mumble a frequency to Bahiri before thumbing a button to continue the search. I don't think it's anything based on what I heard but best to get native speaking Afghani ears on it rather than my cobbled together knowledge of keywords.

We have direct comms to our EW teammate who's embedded with another section, and now and then he alerts us to conversations on other frequencies. Bahiri monitors them too as our patrols move through different parts of the village.

Bahiri and I try not to disrupt the normal flow of the section we are with, but neither of us are infantry soldiers and we know we are an awkward addition with Bahiri weaponless and much closer to me than normal spacing would dictate.

"We're happy to have your awkward arses with us," says our section commander when I remind him he's got us in the middle of his patrol today. "We'll be the first to know what's going on if you guys are in our section."

As the walls of the final compound end, Bahiri and I follow the section out and into the broad, dusty main street. Our spacing widens and among the colourful burkas of the women and embroidered karakul hats of the men, we stalk through them in desert camouflage.

It is market day and stalls are set up either side of the road with vegetables and breads competing for space beside Afghani rugs, pirate DVDs, and knock off sunglasses. The locals have already dialled in how to capture the spending dollars of the international soldiers, but no amount of pirated junk can compete with the dominance of the spicy and sweet fragrance of the markets. Marinades full of garlic and spices drip through the hot grills and sizzle as they hit the coals. The smell of charred meats slow-cooked to the point where they almost fall apart is intoxicating.

Small boys with shiny silver stitching on their waistcoats rush up to soldiers and Bahiri alike. They chatter at a high pitch and swirl en masse like a flock of eager swallows, flapping around one soldier first before peeling off and all swarming around another. The Australian soldiers are limited to repeatedly greeting them or acknowledging their greeting. *Salam Alaikum* or *Walikum a Salam* is the extent of most of their Pashtun. Neither the boys nor the men of Tarin Kowt will speak with me. Even with the helmet, weapon and body armour, they know I am a woman.

Bahiri pushes some money into the palm of a particularly insistent young boy and I hear him give a flurry of instructions. I make out the words goat, hurry and five. As the boy scurries off Bahiri shouts at him to bring back the change.

I know what's about to happen now. Bahiri fluctuates between being worried about his safety and the safety of his family for his role in supporting the Australian forces, and then he defies every protocol that we brief him into that will keep him safe by sending off a local boy to bring back goat kebabs for the section while we are out on patrol.

I try to get the section commander's attention but the boy is too quick. He returns, running out from a side alley and comes skidding back onto the road with his arms full. I see the tension in the shoulders of the soldiers ahead of us. Their casual alertness changes as they stop and prop behind structures and pull their weapons higher into their shoulders. The boy stops dead still in the middle of the road. Subtle changes are enough for the children of Afghanistan to be instantly alert to danger.

"It's okay, it's safe—he's with us," I say projecting my voice but avoiding yelling.

Bahiri joins me. "He has kebab," he says while waving our soldiers off and beckoning the boy to him.

"Kebabs—Bahiri asked him to bring us kebabs," I say loud enough for the soldiers closest to us to convey it to others.

"Lamb kebab—we share one-one," says Bahiri handing a giant roll of roti bread loaded with spiced meat to the rifleman he's been following all day and gestures to the soldier he expects him to share it with.

"*Manana*, thank you," says the rifleman. It is a flavour sensation compared to the ration packs that we eat on patrol, and the kebabs are shared out eagerly. Half the patrol shoves them in their gob as the other half provides security in this relatively safe village that has come to permit us regularly into their market, their hospital and their community spaces.

"Lamb or goat?" I ask Bahiri quietly, thinking I've misheard or got the word wrong from his initial conversation with the boy.

"Goat of course," says Bahiri. "You see sheep here in Afghanistan?"

"Let's just keep that to ourselves then," I say with a conspiratorial wink.

"This is why I tell them lamb," he grins as he offers me the other half of his own delicious meal.

"Thank you," I say as I wolf down the warm bread and meat.

We are soon headed back to the vehicles and onto the FOB. It was a short foot patrol in a tolerant village today in preparation for tomorrow's visitors. Members of the Australian press are being flown in to join us on a vehicle mounted patrol.

〰〰

We drive out the gate of the Forward Operating Base and head to a position above the village we patrolled yesterday.

From inside our command Bushmaster I watch the journos tumble out of the back of the other vehicles and pose for photos dressed up in flak jackets and helmets. When they are ready to talk to the team, our officers steer them away from the soldiers most likely to say something funny, inappropriate or irreverent and towards those they can trust to be as bland as possible. It is important that the message going home to Australian TVs is one of calm control. I hide out of sight so that they don't realise there is a woman on the patrol.

It's not the first time something like this has happened.

When I returned from East Timor I sat on the couch at home and consumed the hours of footage my folks had recorded on their VCR from the news while I was on deployment. I stuffed potato chips in my face and grinned to myself as I watched an interview with two of my very excellent teammates who were based in Dili about their experiences being women on the frontline. They are two totally tough, immensely capable, smart, driven soldiers and EW operators, but at the time they were being interviewed in Dili I was on a chopper to Suai with 1RNZIF nearly 100 kilometres forward of their position. We've all had to do it as EW operators—pretend we are not somewhere, not doing something—or on the flip side, tell a story that you know isn't true because it protects others.

So on that patrol in Afghanistan in 2006 I hid from the journos while they were with us. I'd spend daylight hours inside a vehicle we didn't give them access to and then at night—when I had to go out to go to the bathroom or to sleep—we relied on the body armour and cold weather gear to conceal my shape from them. It did. Didn't fool a single Afghani local. But it fooled every single bloody Australian journo.

Dauntless

AUTHENTICITY
[THE LESSON]

Whether we are concealing our gender, hiding our equipment or covering the source of our intelligence gathering, EW operators have to get comfortable with skating around the edges of the truth. It is the opposite of authenticity, and I have seen the toll it takes on individuals and families to hide parts of themselves and to deal in constant half-truths.

I have been lucky to have my partner do the same job as me, so although we can't talk about anything at home, at work and in secure locations we have been able to debrief an issue or discuss possible solutions together, but this was not the case for many of my colleagues. I saw the walls that they erected between themselves and their loved ones to avoid answering even the most benign questions like "How was your day?" because they didn't want to lie, but couldn't tell them the truth, either.

I saw mates with damaged and broken personal relationships who only looked truly relaxed and open to be themselves when they were with other soldiers. The cover stories and non-truths eroded who they were with their partners, and this guarding of their true selves eroded their relationships and partnerships.

I've seen many soldiers from other trades get caught up in a similar struggle when they leave the ADF too. They find it hard to relate to civilian workmates, other parents at school, or the general public. I hear them say things like "no-one else could possibly understand what it was like, and even when I explain things they still don't get it". So they stop sharing and shield themselves.

This assumption that others 'don't get it' can be unfair. Of course no-one expects someone to understand what it feels like to be in combat who hasn't been, but soldiers forget about the euphemisms they use with each other that have inferred meanings which exclude others from ever understanding.

It's not okay to brag in the Army, so if you've done something cool or well, when telling another soldier you'll allude to it rather than telling them outright. You rely on them filling in the details with their existing knowledge. It's sentences like "I usually take the Minimi," that'll raise an eyebrow from other veterans who know that infers carrying more weight in the weapon and your pack. That it's unusual for a woman to do that. You've got to spell all that out for civilians and it feels a lot like bragging when you have to detail it. Very uncomfortable. So soldiers avoid it, and the void between them and non-veterans gets wider.

I've seen those soldiers retreat into themselves and only come out again when they can interact with other veterans. They feel able to be their authentic selves with other veterans who they feel 'get them'.

It's so difficult for everyone to open up to people who they feel don't 'get them' but when we hide parts of ourselves we hurt our relationships and often, we hurt ourselves. When we can find a way to be truly authentic—absolutely us—something quite amazing happens.

> Today you are you, that is truer than true. There is *no-one* alive who is *Youer* than *You* (Dr Seuss).

I don't mean the 'you' with the perfectly airbrushed pictures, the successful business, and the never-ending stream of awards and achievements. The one who posts photos when they are working from a fabulous beachfront location but never shows the dingy, windowless back office. The version of you who wins, never makes a mistake and who always has a pithy hashtag to position their humblebrag as a cute post, rather than an egocentric boast.

When you are developing your leadership style, it's particularly tempting to act like the version of yourself everyone else seems to be on social media. Don't be.

Be you.

We are bombarded with what we 'should' be as leaders, how we should act, talk, think, and look.

What we 'should' be, is ourselves.

> If you're **curating** you, you're not **being** you.

Dauntless leaders know themselves and know their people. They don't separate their 'work selves' from their 'real selves'. They are not selling an image or faking their competence.

It is not enough to say you are genuine or that you value authenticity. You can't study and then try to demonstrate qualities which are associated with being genuine. Your actions must match your words. You must *be* genuine.

Of course not everyone feels safe enough at work to be their authentic self. Even in the most inclusive organisations people worry that their religion may see them left out of work events, or that they may be discriminated against based on their sexuality. In some workplaces individuals don't feel safe enough to disclose adjustments they might need, or their cultural and linguistic backgrounds. They may not share multiple factors they self-identify with that see them face intersectional barriers and challenges at work.

Dauntless leaders don't shy away from conversations which explore whether people in their team feel safe at work to be their whole and true selves. The best leaders are consistently working on inclusion in their team.

If you can't be you at work, your team will know. Authenticity is essential to leadership because trust matters so much in all our relationships. Our brains can sniff out inauthenticity even when we can't put our fingers on exactly why something stinks.

Human beings are hardwired to detect lying. The micro-lying of inauthenticity seeps into interactions which are picked up by our customers and our co-workers, undermining these relationships.

And don't worry—we've all been there—we've all been inauthentic. Recognising when you're not being you allows you to stop acting and refocus on being. In the times when it is because you fear being your true self, recognising you're not being you is a chance to question whether you want to continue to be in that team, work for that organisation or hangout with this group.

So how do you *be* more genuine?

Being self-aware helps you to be genuine

An easy way to check on your self-awareness and to be more genuine with your team is to consider how you work best. Everyone has different preferences for how they work. You can spend months in a team with someone before you figure out they don't like answering difficult questions before 9am or they prefer email over phone calls.

This was a huge insight we learnt from the Tailored Talent Autism Hiring Program. We originally created an activity for interns to complete about their work preferences in the hope it would assist in making adjustments and supporting self-advocacy, but it was actually a fantastic activity for all our team members, not just those on the program.

It turned out many of the new tools and techniques which helped and supported our interns at work were also incredibly helpful for all our team members. The leaders who took deliberate steps to improve their leadership to be more inclusive and more curious about the needs of their team members in the Tailored Talent program became more inclusive and curious about the needs of themselves and their other team members too.

That first exercise we did with teams was around work preferences and 'some things you should know about me'.

Do you know what yours are?

Here's mine if you need a kick start:

1. Send me an SMS, don't bother calling, I never pick up (if you don't believe me, call me. My voice message says, "I haven't listened to my voice messages since 1945, please send me an SMS". And yes, that's my work number).

2. I work best in the mornings. Issuing me new tasks after 3pm will result in procrastination and poor-quality returns.

3. When I'm working on something deep, I like to stay at it for hours and I will close down my email so I don't get distracted.

4. I'm super into fishing and wakeboarding with my family (I can, and will, bore you for hours about fishing locations, boats and tide times).

When I did this activity with one team I was surprised to see how many people only talked about the things they did like, or how they worked well. They avoided flagging the things they didn't enjoy or they prefer not to do. You'll only get half the result if you only share half of yourself.

> Genuine people own their weaknesses.

Would you share your REAL weaknesses at your next job interview?

Not the 'I'm a perfectionist' rubbish online articles coach you to say—would you share your actual, real weaknesses?

Genuine people are comfortable to talk about their strengths and their flaws because they believe they are good enough.

Not that they are perfect, but that they are good enough, worthy enough, to be themselves. Being comfortable to be open about the stuff you don't do so well is another layer of the veneer so many people put up at work, a veneer you can peel back when you work in a supportive culture and environment.

When you get to this point, you'll also stop clutching at the opportunities which are not a fit for you. When you are honest about your strengths and your weaknesses, you'll find the work which lets you shine and thrive. I see genuine people do this all the time in interviews.

They ask questions like:

"What are the most important skills for someone in this role?"

"What do the top performers in this role do well?"

"What skills are you looking to bring into your team?"

"What gaps are you looking to fill with your next hire?"

They are asking because they want to know what it takes to succeed and they are matching this up with their own strengths and weaknesses in their head.

They don't try to pretend they have all those skills. They talk openly about those they do well, and those they don't, and they ask more questions about those areas of the role they see as unattractive.

I can see them saying to themselves:

"I'll either be a fit, or I won't."

The military helps you do this. Because there is nowhere to hide, your strengths and weaknesses are always known, on display, discussed. The drawback is that you can sometimes put up barriers between yourself and non-veterans for no reason other than you believe they can't or won't understand your strengths, weaknesses, and preferences.

The further I went into my corporate career, the more I learnt to own my weaknesses. Now, when I take on a new project or join a new team, you will never find me talking about 'attention to detail' being a strength. Despite knowing most roles require it and most hiring managers want to hear this phrase, I now refuse to say that I'm good at attention to detail. In fact, nowadays I jump right in and tell future potential leaders straightaway about my disdain for detail.

There is no point in me pretending I like the details stuff (or am good at the details stuff) and securing a job full of work I hate and I'm no good at.

I know what I am good at.

I'm good at strategic thinking, critical thinking and big picture, future orientation. I'm good at leadership, culture and getting stuff done. I'm rubbish at attention to detail, doing things as they've always been done, and completing someone else's plan exactly. I'll always want to make changes.

I focus on my strengths and don't try to hide my weaknesses. I figure either they value the big picture, strategic thinking I can bring, or they don't.

Being candid about my weaknesses has proved beneficial in my work and volunteering. The 'banking and financial' background in my career history has seen me offered several 'treasurer' roles in the community. In reality, I am rubbish at maths and my disdain for details makes me the worst financial controller ever. The organisation gets much more value when I can spend time helping them think about strategic, future-focused goals. And I enjoy it so much more.

The ability to be your authentic self can sometimes be stifled by the team or organisation you work in. Sometimes the inauthenticity barrier for an individual is not the pressure to present as perfect, but rather the way your boss, your workplace or your organisation reacts to challenge or difference of opinion.

Dauntless leaders have a strong enough sense of self to welcome criticism. If somebody criticises one of their ideas, they don't treat this as a personal attack. There's no need for them to jump to conclusions, feel insulted, and start plotting their revenge. They're able to evaluate negative and constructive feedback, accept what works, put it into practice, and leave the rest of it behind without developing hard feelings.

As a leader it's not enough to say you value critical thinking in your team, your actions, and the actions of all leaders in your team must reflect this.

AUTHENTICITY
[THE APPLICATION]

I'd come straight from the Army and, without a single day's experience working in a bank, started as a bank manager in a large regional town.

The organisation had previously promoted a series of 'Bankers' into leadership positions who didn't work out, so they decided to take a risk on someone who had leadership experience and would teach me banking. It was quite a shock to me too, as I'd initially applied for a teller's role as I looked to start out in my post-military career.

I felt like a total imposter and spent the first weeks of my finance career putting on a mask to pretend to be a bank manager. It wasn't just my lack of banking experience. I felt customers asking to see the bank manager were expecting to be greeted by an older man. With a sensible haircut. In a suit.

Not a young woman. With a scruffy ponytail. In boots.

I worried because they weren't seeing what they expected to see, I'd have to act more like a bank manager to make them feel confident I could do the job. So, I changed how I dressed and how I talked. I changed how I acted, and performed my best impression of someone who was stoic, solid, and reserved.

And all it did was create poor relationships with my team and my customers and undermine our trust.

The fake me used language which was stiff. The fake me worked hard to hide my weaknesses and knowledge gaps. The fake me put deliberate space between who I was at work, and who I was at home.

I thought I couldn't be friends with my teammates—hell—I didn't even think I could be friendly with teammates.

Lucky for me, the Global Financial Crisis forced the real me out of hiding quick smart.

I had no choice but to expose my every weakness, trust my team, and to do what was right—what was hard—not what was easy.

≋

"I'm not paying these break costs until you can explain to me exactly how they are calculated," says Mr Jones, our reasonable, but angry customer.

"I can help you, Mr Jones," I say. "Let me just get some details, and we can talk about break costs."

I scurry out of my bank manager's office and into the back-of-house area of the branch.

Behind every teller counter is an area out the back where the safes, the ATM and all the banking paperwork and manuals are hidden. It's always around the corner, tucked away in the bowels of the building, and it's where you can have a conversation out of sight and out of hearing of your customers.

"Etta!" I squeak at a long-standing teller at the branch as I rush through the doors and hide out the back. "Can you help me, please?!"

The desperation in my voice is thick and palpable.

Etta finishes up with her customer and sidles up to me out the back.

"Mr Jones wants to know about his break costs," I gush before I take a big breath. There is no way for me to bluff my way through this I've got to be honest. "What the hell are break costs?"

Etta takes her time. I see a little smile on her lips before she answers.

"It's the difference between the wholesale rate when you applied for the mortgage and when you pay it off," she says as she grabs a brochure and flicks to the right page. "This explains it."

WHHHHHHAAAAAT!!!! I am thinking

I feel like the words she has said are maybe in the wrong order, but I figure if I say them, Mr Jones might know what they mean, even if I don't. I mumble under my breath to try to memorise them.

I clutch the brochure and put my thumb on the equation she has marked. With my shoulders back, I head out to the bank manager's office.

Etta watches me from her teller window.

"Thanks for your patience, Mr Jones," I say. "Break costs are the difference between the wholesale rate when you applied for the mortgage and when you pay it off." I turn the brochure towards him and indicate to the equation.

He looks at the glossy brochure and I peer at it upside down.

There's a backwards 'E' in the equation and lots of super and sub numerals.

Please don't ask more questions, please don't ask more questions, I silently chant in my head.

"Well, what does that all mean?" he says.

Shit, I think. *He's asked more questions.*

"I'll find out for you sir," I say as I gather up the brochure and head out the back again.

"Etta," I hiss. "Can you please help me?"

She comes to me after she's served her next customer with her eyebrow raised.

"What does that all mean?" I ask.

Break costs are complicated to explain to anyone. It's a complex looking equation based on wholesale cash rates. Not the interest rates you pay your mortgage at but the rate the banks buy money at on the wholesale market— so a bunch of data and rates people don't usually see. Etta goes through the details she knows.

It's complex, complicated and ridiculously difficult to try to explain.

I google the equation on a computer out the back and try to understand the maths. I ask every other member of staff who happens to traverse through the back-of-house.

Anything they tell me, I parrot to Mr Jones.

They could have told me to say "purple, monkey, dishwasher" to Mr Jones, and I would have.

I was so out of my depth and so lost. And I was so grateful for their help.

I had no choice but to expose all my inadequacies to every single one of them to try and find a reasonable explanation for our customer.

Together, we found a solution (and a way to dramatically reduce these costs for Mr Jones after an excellent suggestion from one of the team).

〜〜

It was so uncomfortable to have my lack of knowledge exposed and I wondered if my team would ever respect me as a leader again now they knew I knew so little.

What I noticed in the days after the conversation with Mr Jones was the opposite.

My team began to ask me more questions and ask for my help more often. Without planning to, I'd shown them it was okay to not have all the answers. They felt they didn't need to know everything, so were more open about sharing their knowledge gaps and asking each other for help.

I also found out previous managers had avoided conversations with customers which had the potential for conflict. Those leaders who had come before me had made the team have all the difficult discussions and deal with any angry customers. They didn't step in.

I didn't know much about banking, but I sure knew if things were getting difficult with a customer, I would have wanted my boss to help out. So I stepped in—even though I had no idea what the conversation was about most of the time.

My team appreciated my willingness to take on the difficult conversations and they were happy to provide the technical expertise, because I was providing the top cover.

The team showed me that my acting a role, instead of being myself, was getting in the way of us really knowing each other and working together as a team. I started to notice how my behaviour shaped the way my team behaved in multiple scenarios.

This first example—when exposing my weaknesses resulted in everyone feeling more comfortable asking for help—made me self-aware of my other behaviours, and the impacts (both positive and negative) on how my team would then behave.

Once I stopped pretending to be a bank manager and started being me, I was way more flawed, but I built much deeper connections with everyone around me.

I could feel my team, and our customers, making space for me to learn, grow and thrive.

Once I stopped putting up the walls, others bought theirs down too.

ACTIONS—AUTHENTICITY

REFLECT

How much time and effort do you put into hiding your weaknesses?

- Do you feel that you can be your authentic self at work? How about at home?

- How do you think your team would answer those questions?

- What do you do as a leader to actively encourage authenticity and inclusion?

GET CURIOUS

Ask three team members:
- What do you think we do to promote inclusion in our team?

- What stops people in our team from being their authentic selves?

- When do you feel like you can really be yourself and be accepted for who you are? work/home/sport/ other/ never?

TAKE ACTION

Get right to the heart of each team member's work style with this quick exercise in your next team meeting.
- Have everyone in the team fill in their work preferences
  AM or PM or other – why?
  Communications preferences?
  How do you work best?
  What stops you working best?
  What else is important to you?

- Have pairs 'swap cards' and discuss their preferences for 5 minutes

- Form new pairs and repeat

- Encourage team members to share cards at the start of any new project, in any new sub-teams they form and/or keep them on a shared drive.

REFLECTIONS ON COURAGE

I learned a lot about courage in the Army.

A lot about finding my own courage and a lot about courageous leadership. Not the gungho bravery type of courage. I mean the courage to stand up and offer a different view. The courage to be ready to change and grow. The courage to take on something new. The courage to stand up for others and encourage them.

Dauntless leaders find a way to be okay with the uncomfortable. They know that this is where they grow and make change. They know that even though sharing your weaknesses can feel uncomfortable it will ultimately lead to better relationships and opportunities.

All the best leaders I know use this 'uncomfortable zone' to hone their courage. They use the discomfort to change their world. Whistleblowing, lifting others, championing inclusion, being their authentic self; all of these take personal courage and a belief you can push through and achieve more than others think is possible.

COURAGE—where to now....

REFLECT

Review your answers from the start of the courage section. Consider:

- How comfortable are your team with change?

- How diverse is your team? Consider diversity of thought, ability, background, culture, sexuality, gender. Does your team reflect the community you represent or customers you serve?

- What do you believe your team share or hide?

GET CURIOUS

Ask three people in your team about courage:

- How do they feel about change?

- Who do they advocate for? Mentor? Promote?

- What do they not do/talk about at work that they do/talk about at home?

TAKE ACTION

Talk to your team collectively about courageous behaviours. What stories about growth, advocacy and authenticity already exist in your team? Agree how you can champion these stories and identify an action you can take in each of the three courage areas in your team.

Share something genuine and authentic about yourself with your team that you've been holding back on. Talk with them about why you were holding back.

Take a deliberate approach to mentoring. Who is someone each team member can mentor that will change the game for both the mentee and mentor? Agree when everyone will set up their first meetings.

DAUNTLESS LEADERS

As a leader, you cast a long shadow.

What you do, why you do it and how you do it will all have broad and long lasting impacts on the people you lead.

Those people brave enough to do the right things, not the easy things, during their time leading a team will need some dauntless-ness about them.

It requires determination to keep focusing on culture instead of pushing for this week's numbers.

It takes tenacity to remain team-oriented and not self-centric.

It demands nerve to courageously stand up for others and yourself.

You will need to be Dauntless if you want to cast a shadow with an impact worth leaving.

Dauntless leaders build great CULTURE

From nothing, from groups of strangers, with little time.

They know teams matter when it comes to the crunch, so they put all their effort into building team culture and an environment where people feel safe to speak their mind.

They tell the truth; ugly, direct, uncomfortable or confronting.

They communicate with clarity and trust their people to do their jobs well.

They give their power away.

Dauntless leaders focus on TEAMS

They encourage different thinking and understand inclusion is the key to harnessing diversity of experience and thought.

They help people leverage their strengths and find purpose in their work.

They encourage behaviours which align with the teams values and create opportunities for their team to put these values into practice.

They acknowledge when people do the right thing and behave in the right way.

They acknowledge when people do the wrong thing and behave in the wrong way.

They are right there in the thick of it with their team— sharing experiences, rolling up their sleeves, and providing support when things get hard.

Dauntless leaders show COURAGE

They do what is right, not what is easy.

They embrace risk and have a growth mindset.

They back their team and they lift others up.

They are authentically themselves—the good, the bad and the ugly.

~~

Afghanistan seems like such a world away for me now. Sometimes it feels almost like it happened to a different person but then I remember the pieces that still apply in this new world I live in. The new corporate me and the old Army me is a blend of both the people I've been and the person I am now.

I have no doubt you have these different versions of yourself too. One you might lean on when you need technical expertise. Another aspect might come to the fore in a crisis or under time pressure. When you can reflect on what you learnt about yourself, your leaders and leadership during those times, you'll be able to follow those individual strings in your own ball of tangled up and, interconnected pieces of information, and follow them through to some incredible insights.

Being more Dauntless will take work, but it will be worth it through the impact that you make.

Wherever you go from here, I'd encourage you to take some sort of action. In the military we often say 'done is better than perfect'. I see too many people waiting for the perfect time, or the perfect opportunity. The time is now, the opportunity is everywhere.

If you put this book to the side and do nothing, you'll gain nothing. So, if you picked it up because you wanted to learn something about yourself or change something about your leadership style, then I'd encourage you to do something— **anything!** To lean on a more Army phrasing—you've read the book, so what are you going to fucking do about it?

Pops chin, gestures with knife chopping hands, walks off to demand more of the next person

THANK YOU

I firmly believe no-one achieves anything alone.

I owe enormous thanks first to Damo and Will. I'm very lucky to have two such great blokes as my family. Thank you for helping me to remember all my stories, encouraging me to follow my wildest whims and always being in my corner.

I've got a massive, extended pile of friends who are family and family who are friends. I won't name you all, but thank you all for the laughs, the hangs and for being there when everything is awesome *and* when everything is shit. Thanks for letting me write at your houses, swim in your pools, drink your beers and talk a whole load of shit with you all.

To the many Dauntless leaders who are featured, alluded to or relegated to the document called 'CUTS' for this book, thank you for your leadership. Thank you for supporting and encouraging your people to be their best. Thank you for choosing to do what is right, and not what is easy. Thank you for all the opportunities you gave me to grow, learn and develop.

Thank you to the incredible Beta readers who made such a massive difference to this book. Damo, Kris, Neal, Katie, Graz, Catherina, Chris, Bek, the other Bek, Glenn, Aaron and Nic—your words made an enormous difference to my words. Thank you.

Thanks to the fantastic crew who helped bring this book and brand to life. Cat, Hammad, Rach, Kim and the Change

Empire Books team, none of this would exist without you. To Shae, what an absolute knockout picture and to Lis, for beautiful design and advice.

I've been lucky in my varied careers to work with so many passionate leaders and supporters. Many people have given me a leg up, a chance to try something new, or an opportunity to forge a new path. I hope I've delivered on the faith you've all shown in me and hope you all continue to open doors for others.